The STURM, RUGER 10/22 RIFLE AND .44 MAGNUM CARBINE

Duncan Long

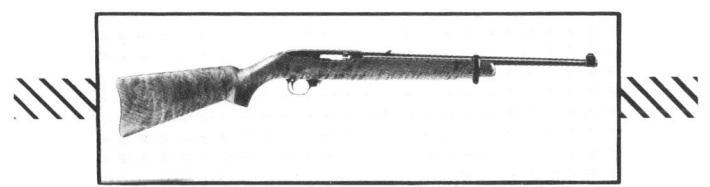

Paladin Press
Boulder, Colorado

Also by Duncan Long:

AK47: The Complete Kalashnikov Family of Assault Rifles
AR-7 Super Systems
The AR-15/M16: A Practical Guide
AR-15/M16 Super Systems
Automatics: Fast Firepower, Tactical Superiority
Combat Ammo of the 21st Century
Combat Revolvers: The Best (and Worst) Modern Wheelguns
Combat Rifles of the 21st Century: Futuristic Firearms for Tomorrow's
    Battlefields
Homemade Ammo: How to Make It,
    How to Reload It, How to Cache It
Making Your AR-15 into a Legal Pistol
The Mini-14: The Plinker, Hunter, Assault, and Everything Else Rifle
Mini-14 Super Systems
Modern Ballistic Armor: Clothing, Bomb Blankets, Shields, Vehicle Protection . . .
    Everything You Need to Know
Modern Sniper Rifles
The Poor Man's Fort Knox: Home Security with Inexpensive Safes
Powerhouse Pistols: The Colt 1911 and Browning Hi-Power Sourcebook
The Ruger .22 Automatic Pistol: Standard/Mark I/Mark II Series
Streetsweepers: The Complete Book of Combat Shotguns
Super Shotguns: How to Make Your Shotgun into a Do-Everything Weapon
The Terrifying Three: Uzi, Ingram, and Intratec Weapons Families

*The Sturm, Ruger 10/22 Rifle
    and .44 Magnum Carbine*
by Duncan Long
Copyright © 1988 by Duncan Long

ISBN 0-87364-449-2

Printed in the United States of America

Published by Paladin Press, a division of
Paladin Enterprises, Inc., P.O. Box 1307,
Boulder, Colorado 80306, USA.
(303) 443-7250

Direct inquiries and/or orders to the above address.

PALADIN, PALADIN PRESS, and the "horse head" design
are trademarks belonging to Paladin Enterprises and
registered in United States Patent and Trademark Office.

All rights reserved. Except for use in a review, no
portion of this book may be reproduced in any form
without the express written permission of the publisher.

Neither the author nor the publisher assumes
any responsibility for the use or misuse of
information contained in this book.

Contents

Chapter 1
The Ruger 10/22 & .44 Carbine   1

Chapter 2
Variations   13

Chapter 3
Care & Maintenance   29

Chapter 4
Accessories   57

Chapter 5
Ammunition   87

Appendix A
Manufacturers   95

Appendix B
Publications & Videotapes   101

Warning

Technical data presented here, particularly technical data on ammunition and the use, adjustment, and alteration of firearms, inevitably reflects the author's individual beliefs and experience with particular firearms, equipment, and components under specific circumstances that the reader cannot duplicate exactly. The information in this book, therefore, should be used for guidance only and approached with great caution. Neither the author nor the publisher assumes any responsibility for the use or misuse of information contained in this book.

Acknowledgments

Special thanks to Linda M. DeProfio of Sturm, Ruger and Company, Inc. for her help in securing information, background history, and photos of the Ruger 10/22. Thanks also to officials at Sturm, Ruger and Company, who graciously loaned firearms to me to test out during the writing of this and other books, as well as the many other companies that sent sample products to inspect and test.

Thanks also to the hard-working and skilled people at Paladin Press who are able to work the magic of turning a jumbled pile of typo-filled manuscript pages and dog-eared photos into a real, live book.

Finally, thanks to Maggie, Kristen, and Nicholas for making yet another book possible.

Chapter 1

The Ruger 10/22 & .44 Carbine

The Ruger 10/22 has become one of the most popular .22 rifles in the world, and not by accident. The 10/22 is designed for a long useful life. If and when parts do finally wear out, they are easily replaced or repaired, thanks to a design that makes the rifle easy to disassemble by a gunsmith or even by the owner of the firearm.

Too, there are aftermarket accessories galore. Entire companies do little more than stamp out accessories for the 10/22 that transform it into a lightweight camping gun, a compact survival rifle, or even a quasi-assault rifle that is considerably cheaper to plink with than its centerfire brothers.

The gun's reliability is another important feature. Many firearms are introduced to the public with a few bugs in them and generally function poorly until the owners either take them to gunsmiths for tuning or buy the model a year or two after release, when the manufacturer has finally worked the faults out of the design. The 10/22, however, hit the marketplace thoroughly debugged. Sturm, Ruger and Company designers did their homework before marketing the rifle, and this has certainly helped its image from the very beginning.

As an example of just how reliable the 10/22 is, noted firearms writer and gunsmith J.B. Wood wrote in his book, *Troubleshooting Your Rifle and Shotgun:*

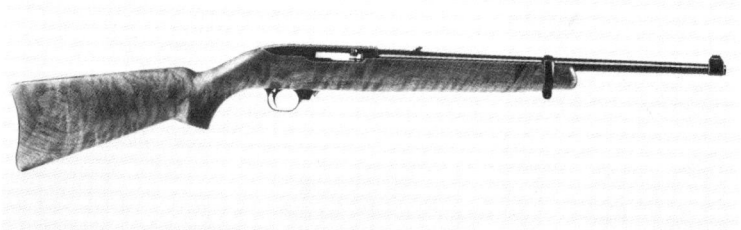

One of the most popular .22 rifles in the world, the 10/22 is designed for long use. If and when parts finally do wear out they are easily replaced or repaired, thanks to a design that makes the rifle easy to disassemble by a gunsmith or even the owner of the firearm. (Photo courtesy of Sturm, Ruger and Company)

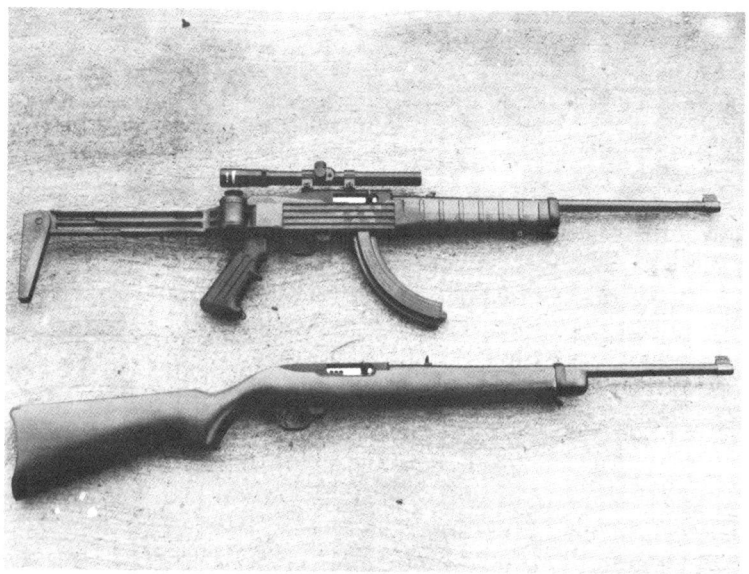

The 10/22 is a very reliable .22 rifle for which a wealth of aftermarket accessories is available. Owners of the 10/22 can quickly transform a standard model (like the lower carbine) into a high-tech "combat-style" firearm without any special know-how or gunsmithing work.

... As a subject for "troubleshooting," this one [the 10/22] is going to be very difficult, since hardly anything ever goes wrong with it. The cartridge feed system is particularly outstanding, and a 10/22 that jams is almost unheard of. In fact, in the past 14 years, I've not encountered a single one. I can even recall one case in which the gun had a ruined sear as a result of amateur tampering, and this one was firing full auto. It would empty the magazine in a single burst before the finger could be removed from the trigger, but it didn't jam!

The 10/22 came into being through the back door, as it were, since it was a spinoff of Ruger's .44 Magnum Carbine which, in turn, undoubtedly was greatly influenced by the U.S. M1 Carbine.

The M1 Carbine never proved to be exactly what anyone expected it to be. Work on the weapon began in mid 1941; the final design was selected from eleven different rifle entries. After extensive testing of the prototypes, the carbine designed by David M. "Carbine" Williams was selected because it was the most reliable and easiest to make.

On paper, the M1 Carbine was designated for use by military personnel who otherwise would have used the .45 pistol; when actually deployed in the field, the M1 Carbine proved so popular, because of its handy size and light weight, that it was soon being issued to a much wider range of troops.

While the M1 Carbine was popular with the troops, its performance was far from ideal; its takedown was rather complicated and it was saddled with a cartridge that was, at best, marginal even at close ranges. Other problems included less than ideal magazine release and safety/selector positions, a fragile rear sight, receiver guide rails that often broke with extended use, and extended magazines that often failed to feed properly, especially in the automatic mode.

Following the war, many returning veterans tried to use the M1 Carbine for hunting deer and other medium-sized game. Again, the cartridge proved to be marginal even with modern expanding bullets. Word soon got around that the M1 Carbine was useful only for plinking, but many hunters still longed for a lightweight, semiauto carbine that could be used for deer hunting and the like. This was the market for which Ruger's .44 Carbine was developed. But there's more to the story than that. To properly understand how the .44 Carbine came about, it's necessary to look at the man at the helm of Sturm, Ruger and Company.

William Batterman Ruger has proven to be one of the major firearms geniuses of the twentieth century. Born in 1916 and raised in Brooklyn, New York, Ruger and his father enjoyed the outdoors and often went on hunting trips. Ruger also spent many summers at his grandfather's, where he often found a chance to shoot the .22 rifle that had been his twelfth-birthday present. This may explain why Ruger's first successful firearms were .22s. The fact that his grandfather was a painter may also explain the artistic bent Ruger shows in creating arms that have consistently enjoyed classic and clean styling as well as good design over the years.

Like many other success stories, Ruger's didn't happen overnight. His business seems to have developed from an early interest in both firearms and engineering; as a boy, Ruger sifted through all the books he could find on the subjects in the New York City Library.

One of Ruger's first firearm designs was created during his two-year stint at the University of North Carolina. Toiling in a basement (when many probably thought he should be doing homework), Ruger perfected a semiauto conversion of a Savage 99 lever-action rifle. Later, he wrote about the conversion in *American Rifleman* magazine.

Using $10,000 he inherited in 1939, Ruger dropped out of college and took a short trip to Europe with his new bride. Upon returning to the United States, Ruger started looking for a job after settling down in Hartford, Connecticut.

Ruger tried to work for, or sell ideas to, a number of firearms manufacturers, including Savage. None of his efforts to work for the companies panned out, however, and Ruger soon ended up working in a small machine shop in Greensboro, North Carolina, for twenty dollars a week. It was undoubtedly during this period that Ruger started to see various ways to take advantage of modern industrial methods and machinery to create gun designs that could be built with less labor and machine work, thereby keeping costs down.

Shortly after this stint as a machinist, Ruger again tried his hand at firearms work with several prototypes he had created. These led to his being hired as a gun designer for the U.S. Army at Springfield Armory in 1939. Things didn't go smoothly for Ruger. Undoubtedly the team effort and possibly the lack of interest in innovation among his military overseers quickly disillusioned him. After less than a year, Ruger quit.

Ruger again found himself in North Carolina, making parts for a light machine gun he had designed. Ruger was able to show a working model of his new weapon to several manufacturers. One company, the Auto Ordnance Corporation (which made the Thompson submachine gun), showed interest in the design and hired Ruger in 1941 to help develop and demonstrate his new gun, which he had modified to conform to the U.S. War Department's machine gun requirements. The U.S. Military, however, was continuously changing requirements for its weapons so that, while Ruger's prototype was tested at the Aberdeen Proving Ground, it was never accepted.

In 1946, Ruger quit Auto Ordnance and opened his own small manufacturing business. Thinking that the economy would be on an upswing following World War II, Ruger decided to manufacture quality carpenter's tools. He located his small business in Southport, Connecticut. Again, however, success did not seem to be in the cards. Though the tools he produced were of high quality, the demand wasn't there. Sales weren't good and by 1948 the company was out of business.

Undoubtedly many people would have given up at this point, but Ruger was learning that products must have a wide market to succeed and also must be able to be made inexpensively without sacrificing quality. With both his manufacturing techniques and business lessons learned the hard way, Ruger probably would have eventually succeeded in creating a thriving gun manufacturing business, but a bit of luck came his way to speed things up. Ruger made friends with a neighbor named Alexander M. Sturm.

Sturm was a painter who collected guns for a hobby; this hobby led to Sturm's interest in a new .22 pistol that Ruger had designed. One thing followed another and soon, with Sturm putting up most of the money to finance the new business, Sturm, Ruger and Company was formed to manufacture the new gun.

In January 1949, Ruger started creating tools to make the new pistol, as well as handling the paperwork, making preliminary sketches and production drawings, and doing other administrative tasks. After the tools were created to build the new gun, a staff of twelve helped Ruger turn out parts in runs of one thousand at a time. By fall, enough parts had been made to assemble one thousand pistols and the gun was introduced for sale through the mail, which was perfectly legal back in those good old days.

A small magazine ad, coupled with a very favorable review by noted gun expert Major General Julian Hatcher in the "Dope Bag" section of *American Rifleman*, launched the business. Orders came pouring in for this low-priced, high-quality pistol, which had almost no competition in the marketplace.

The new Ruger semiauto pistol was low in price because its design took advantage of manufacturing techniques that minimized costly hand work and milling of steel parts. The Ruger pistol sold for only $37.50; the only similar firearm was the Colt Woodsman, which sold for $60.

Shortly after this, Sturm died; the company's falcon trademark, which Sturm had designed, was changed from red to black on the .22 pistols to commemorate his death. Despite Sturm's death, the company continued to grow, with Ruger carefully managing it each step of the way. By 1953, Sturm, Ruger and Company was well established.

At this point, Ruger introduced a new single-action revolver that he had designed. The timing could not have been better, since Hollywood was cranking out Western movies and TV shows and the fast-draw craze was sweeping through the shooting public. Although demand for single-action pistols had mushroomed, the old Colt single-action pistol everyone wanted was out of production. Ruger's new pistol satisfied a marketplace demand and carried a low price tag.

The new Ruger handgun was an improved, updated version of the old single-action iron. Not only was the new revolver less expensive and tougher than other single-action guns, it also was chambered for the inexpensive .22 LR, making it cheap to plink with. Word soon got around that the Ruger gun could take anything the user dealt out and the new single actions sold like the proverbial hot cakes. Later, a somewhat heavier single action chambered for centerfire cartridges was added to the Ruger lineup and captured even more of the single-action market. This heavier gun could be used for handgun hunting or even self-defense.

By 1959, Sturm, Ruger and Company had outgrown its original plant. After a move to a larger building, the next year Ruger had to build an addition to the plant to make it 50 percent larger in order to handle all the orders being received.

Companies that are under the control of one savvy designer often fail to grow because the person who created them is spread too thin trying to manage the business. There is no time to create new products or, often, even keep on top of marketing and managing problems. Bill Ruger proved to be one manager who didn't get bogged down. There are four reasons for this.

One is that Ruger seems to be able to market firearms that sell. His marketing style seems to be purely the seat-of-the-pants method frowned upon by modern business schools. Ruger has never used market studies; he relies on his own good judgment. Thus far, Ruger has proven to be a great judge of things and has tended to set trends rather than follow them.

As Ruger once told an interviewer:

> When manufacturers go around making surveys and asking people what they want, what they're really saying is that they don't understand the business they're in. Our business is more than a business, it is like some sports--it has a heart. I mean, guns aren't just something you make like tools or chairs or some other utility objects. Guns are valued possessions and provoke all sorts of emotional responses in people. If the manufacturer doesn't know what makes a really appealing gun, then he isn't going to have much success. I can't say I know all about it--I wish I could--but I do love guns, and that's been a big help.

Ruger's second skill is a knack for forming teams of designers and workers who can perfect designs and debug new firearms. This allows the company to produce and market a wide variety of firearms that, when they hit the marketplace, function with peak reliability right from the box.

Third, Ruger manages to keep the "classic" outline on all the firearms his company produces. This is an important marketing point, since most shooters are concerned about more than just whether or not a firearm functions well. The look of the firearm is important, perhaps more important than most gun owners will readily admit. Guns that complement their functional lines with eye-pleasing shapes sell better than ugly ducklings that may actually outperform the lookers. Here again, Ruger's guns outshine much of the competition.

Finally, all the new firearms designs that Sturm, Ruger and Company produces are representative of Ruger's technique of coupling the design to modern industrial methods. These manufacturing techniques make many of the parts both tougher than those made by older methods and less expensive to boot. The result is a more rugged firearm that carries a lower price tag than competitive guns.

In 1960, shortly after the move to the new plant, the company started marketing its first rifle: the .44 Magnum Carbine. This gun combined many features found on the M1 Carbine with newer styling as well as the classic lines (especially the buttplate) of an Old West saddle gun or a Sharps rifle. Perhaps most conducive to a slick look is the fact that the gas-operated rifle used a tubular magazine rather than a detachable box. With the small magazine capacity of a hunting rifle, the tube on the .44 Magnum Carbine doesn't extend past the front of the stock and the rifle's lines aren't marred by an unsightly magazine hanging from its lower edge. The new rifle was first marketed as the Deerstalker. Unfortunately, Ithaca was selling a Deerslayer shotgun at the time, which caused not a little confusion among the shooting public. Shortly after release, the rifle became known simply as the Ruger .44 Carbine.

While the M1 Carbine had been a failure as a hunting rifle, the new .44 Carbine, firing the powerful .44 Magnum (which had ballistics similar to those of the .30-30), proved to be an ideal short-range weapon with the low recoil enjoyed by many gas-operated rifles.

By 1963, Sturm, Ruger and Company was producing so many firearms that it was time to expand again. A subsidiary company, Pine Tree Castings, was formed to handle investment castings. The Pine Tree plant was completed in 1964 and is located in Newport, New Hampshire. Shortly after opening, Pine Tree Castings became the largest producer of investment castings for firearms in the United States, since Sturm, Ruger and Company uses more castings in its firearms than the rest of the firearms industry combined. By 1984, Pine Tree had grown to twenty-two times its original size. A second division, Uni-Cast, at Manchester, New Hampshire, has since been created for casting gun parts for Ruger and other companies.

Investment casting, also known as the "lost wax process," starts by creating wax models of the final part from relatively inexpensive molds and then placing fine, sand-like material around them. These pieces are then baked to melt out the original wax model and steel is poured into the sand molds formed around the wax. After the steel hardens, the sand is removed; the part is ready to use after a minimum of machining to bring it to final shape and size. (For an excellent demonstration of how this process works, as well as an interesting look at Bill Ruger and his company, see the 60-minute videocassette, *Conversations With Bill Ruger*, available from the Blacksmith Corporation for about $60.)

The real advantage of investment castings is that the wax model used to create them can be very detailed, so that the final steel casting is very nearly finished rather than roughly sized, as is the case when steel bar stock is used. An additional plus is that the crystalline structure of the steel tends to be more complete in the investment casting, so that it is stronger than the same part created by machining. Machining is necessary on investment casting pieces only when surfaces must be very smooth or polished; even then, the task can be carried out with a minimum of work.

Because investment cast parts are inherently stronger and can even be cast using steel alloys that aren't readily machined, Ruger guns using such parts are considerably stronger than similar machined firearms without adding to their weight or resorting to the use of welded sheet metal, plastics, or aluminum. As an example of how much stronger an investment casting can be, the machined bolts of the Springfield '03-A3 rifle and the Mauser 98 will fail under fifteen thousand and eighteen thousand pounds of pressure, respectively, while the Ruger M-77 bolt, basically identical except for being an investment casting, will not fail at up to twenty-five thousand pounds.

Strangely enough, the .44 Carbine had only six parts made of investment castings; for some reason the receiver was machined from a solid bar of steel. Why Ruger departed from his usual method of using investment castings for such parts is unknown, but the machined receiver of the .44 Carbine is probably the main factor that led to its demise. The Ruger .44 was followed by a sister rifle in 1964. The new rifle was chambered for the .22 LR and had a detachable box magazine holding ten rounds; while the .22 had only two major parts made of investment castings, a number were created of stamped metal and/or easily machined aluminum. The magazine capacity of the new rifle, plus its caliber, led to its name--the 10/22.

The 10/22 was unlike any other .22 rifle being marketed in 1964. It could function reliably with both standard velocity and the then-new high-velocity .22 rounds. It did so almost without failure, thanks to a design that retarded cycling of the bolt slightly to give new rounds time to position themselves on the magazine.

When first introduced, the 10/22 carbines were billed as being similar to the .44 Carbine. The new 10/22 looked like the larger-caliber rifle even though the two had dissimilar magazines; this was achieved by designing the 10/22's magazine to fit flush with the lower side of the stock. Making a box magazine with a large capacity that could do this took a design team eighteen long months to perfect.

Apparently the basis of the ten-round magazine was an idea that had been bouncing about in Bill Ruger's skull for some time before the 10/22 was created. In 1964, Warren Page wrote in the July issue of *Field And Stream:*

> I happen to know Bill Ruger rather well and for years have listened to him grumble that the finest feature of the original Mannlicher action, the grease-slick rotary feed, had never been adapted to a U.S. sporting bolt action.

One reason for this was complexity and expense. The design of the Mannlicher action called for a lot of machining and had to be made of a material that would be more or less self-lubricating, since most lubricants can easily deactivate ammunition. While the Savage Model 99 (as well as a few other firearms) did create successful variations of this action, they were expensive to make.

Then Ruger's design teams tackled the problem. Their solution was to place a steel feed lip at the top of the magazine where friction would wear out a lesser material, and use a tough, slick plastic for the magazine shell and internal rotary element. The nylon-like plastic, carrying the trade name "Celcon," was both strong and self-lubricating. While the task wasn't easy, the result was a strong, compact magazine that was inexpensive to manufacture and functioned flawlessly in the field.

Despite the differences in the magazines, feed systems, and modes of operation, the 10/22 and .44 Carbine were very much alike in many aspects. Both rifles used trigger groups that could be removed easily in block by drifting out the two pins holding them to the receiver. In addition to a similar fieldstripping procedure, the two guns had identical sighting systems, safety positioning, charging handles, and overall lengths. About the only difference in feel was in weight; the 10/22 weighed 5 pounds while its centerfire counterpart was 5 3/4 pounds.

Since .44 Magnum shells are expensive for practice, early Ruger ads played up the fact that the 10/22 was the spitting image of the .44 Carbine. One of the ads that appeared in 1965 told readers that, "These two Rugers are ideal hunting companions . . . identical in size, balance and style and nearly the same in weight. Sharpen your shooting with the 10/22--and you'll be in peak shooting form when you go after brush-country deer with your .44 Magnum Carbine . . ." But, perhaps to Ruger's surprise, the 10/22 actually reached a much larger market. The new .22 proved to be both reliable and accurate, and became popular in its own right, finally outlasting the .44 Carbine, which was discontinued in 1985.

In 1984, Arcadia Machine & Tool (AMT) cut into some of the Ruger market by marketing two new stainless steel guns under the Lightning trademark; one was a copy of the Mark I .22 pistol and the other, a copy of the 10/22 carbine (such copies were legal to make because patent rights had run out on the two firearms). The AMT Lightning pistol is different from the Ruger Mark I only in several cosmetic changes; an adjustable trigger and scope-mount cuts were added to the .22 pistol.

The Lightning 25/22 rifle's main selling point is that it's made of stainless steel. It's available with either a fiberglass stock or with a Choate folding stock; its receiver is dovetailed for scope mounting and it has an extended magazine release. The gun's name comes from the fact that it uses an extended "banana" magazine that holds 25 rounds rather than the flush mounted 10-round magazine used by Ruger.

While the AMT Lightning pistol and rifle work well, it's generally felt that the original Ruger weapons outperform the copies. That, coupled with the higher price tags of the AMT guns, has probably kept AMT from cutting into Ruger's market to any great extent.

The Ruger 10/22 has proven to be one of the most popular .22 rifles of all time, with owners ranging from the plinker to the casual target shooter to the person needing a survival rifle to carry in an airplane or stock in a fallout shelter. Such popularity has led to aftermarket businesses that offer a wealth of good-- and not so good--accessories for the 10/22. These, in turn, create more buyers for the 10/22 as people discover that the rifle can be quickly customized. The 10/22 promises to remain popular for some time. Its accuracy and reliability, coupled with a wealth of accessories and a low price tag, make it first choice as a .22 semiauto rifle.

Chapter 2

Variations

The designs of the 10/22 and .44 Carbine make both of these firearms very easy and enjoyable to use.

A cross-bolt safety is located in front of the trigger guard where it is snapped off easily (by right-handers, at least) with a flick of the finger. When this is done, a red ring is exposed on the left side of the safety. Pushing the button from left to right again returns it to safe.

Left-handed shooters may find this setup less than ideal. Several options are available that might enhance the use of the rifle by left-handed shooters. One would be to have a gunsmith make a reverse safety so that a push from left to right places it in the fire position; another option would be placing an M1 Carbine-style lever in the safety hole. Either solution would entail a bit of work and expense, however, and most shooters will prefer to put up with the very slight inconvenience of the original setup. Interestingly enough, Bill Ruger is left-handed. Apparently he doesn't see the 10/22 safety arrangement as a shortcoming or he undoubtedly would have done something about it.

Both left- and right-handed shooters will find the rest of the rifle easy to use. The charging handle is easily pulled back with either hand; the bolt hold-open lock (located in front of the trigger guard) and magazine release (to the rear of the magazine well) are more or less ambidextrous.

A cross-bolt safety is located in the front of the trigger guard and the bolt hold-open lock is located in front of the trigger guard. The magazine release is just to the rear of the magazine well and fits flush with the lower side of the stock.

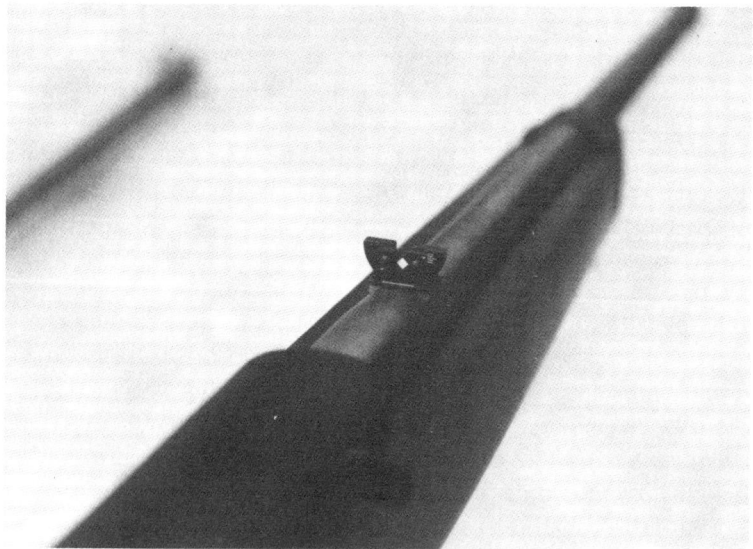

The rear sight on the 10/22 and most Model .44 Carbines gives the user a choice of two apertures. One side of the sight blade is a rounded notch while the opposite side is a square. To switch from one to the other, you must remove the two screws holding the blade, turn the blade over and remount it, and replace and tighten down the two screws.

The magazine release is easily worked by pushing against it at the rear of the magazine with the thumb while tugging at the magazine with the thumb and forefinger. With aftermarket extended magazines, the operation is even easier. Although the bolt doesn't lock open after the last shell has been fired, it can be locked back for easy cleaning or inspection. To do this, the bolt is pulled back and the hold-open locking lever at the outside front of the trigger guard is pulled back and downward. To release it, the bolt is pulled back and the lever pushed back and upward.

The rear sight is a bit fragile and should be folded down when the rifle is being disassembled or is not in use. The sight is mounted in a milled cut in the barrel and can be drift-adjusted for windage or removed. For elevation adjustments, two screws on either side of the rear sight blade must be loosened, the blade moved up or down, and the screws tightened. Fortunately the blade has white markings so the shooter won't lose track of where the blade was set--or where it ends up. Ruger quality control is such that the sights are generally pretty well zeroed at the factory; many shooters never have to adjust them.

The rear sight also gives the choice of two apertures. One side of the sight blade is rounded, while the other, on the opposite side, is a square notch. To switch from one to the other, remove the two screws holding the blade, turn the blade over and remount it, and replace and tighten down the two screws.

The front sight is a modified ramp with a gold bead. The front sight blade is placed in a cut on the muzzle band of the barrel and can be easily replaced or drifted for windage.

The top of the receiver is drilled for a scope base; the base and mounting screws are included with new rifles. Although the base is designed for use with 3/4-inch .22 scopes, many shooters prefer to use 1-inch scope adapter rings. This arrangement allows shooters to take better advantage of the 10/22's accuracy and reduce the target-acquisition time, due to the wider field of view of the larger scopes.

At one time, a tip-off mount adapter was offered for the 10/22 and the .44 Carbine. This allowed a scope to be rotated out of the way so that iron sights could be quickly used if the scope were damaged or became fogged. At the time of this writing, these do not appear in the Sturm, Ruger and Company catalog.

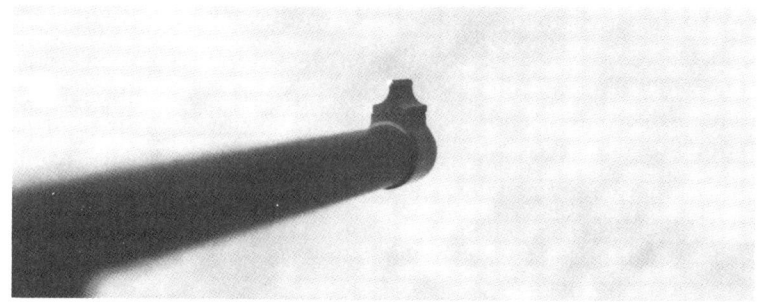

Late-model .44 Carbines and current 10/22s sport an excellent gold-bead front sight. The blade is placed in a cut on the muzzle band of the barrel and can be easily replaced or drifted for windage.

The top of the receiver on the 10/22 is drilled for mounting a scope base; the base and mounting screws are included with new rifles. The rear sight can be flipped down when the scope is in use.

Variations

The 10/22 stock, as with most carbines, is a bit short. Most shooters don't notice this after using the 10/22 for a while, because the carbine's light weight and short barrel counterbalance the shorter pull length.

Internally, Ruger has slipped in a number of innovations, with several parts in the trigger group of the 10/22 doing double duty. Perhaps most interesting is the way the cycling rate of the rifle has been reduced so that rounds have time to feed up to the lip of the magazine before the bolt strips each one off and chambers it.

Two features slow down the bolt's rearward movement. The first is the use of an internal hammer, rather than a striker, to drive the firing pin into the cartridge. During recoil, the hammer has to be forced back, and this action retards the bolt's initial backward travel. At the end of its backward cycling, the bolt is further retarded by a steel bolt-stop pin that runs across the back of the receiver. When the bolt hits this pin, the curved rear of the bolt cams it downward slightly; this motion eats up a bit more of the recoil energy. The bolt hesitates at the rear of the receiver for a moment before the recoil spring pushes it forward slightly to straighten out before traveling forward to chamber another round. In addition to aiding the chambering of the next round, this retardation of the cycling rate also enables the bolt to be light and the single recoil spring to be rather small.

The 10/22's bolt, barrel, and trigger group are made of steel; weight has been shaved off the gun by making the trigger housing and receiver of aluminum. The steel barrel is joined to the aluminum frame by a V-shaped bar that dovetails into a cut in the lower side of the barrel, which is then screwed into the receiver with two bolts. This makes it easy to remove the barrel with a hex wrench (or a screwdriver on early models) while assuring that the barrel is properly aligned when replaced.

The 10/22's magazine consists of seven different parts. The lips, which are exposed to heavy wear, are made of steel, while parts that need to be self-lubricating are made of plastic. A coil spring placed inside the magazine rotor rotates shells upward to be stripped off the feed lips.

When 10/22 was first marketed, many felt that its magazine should be completely stripped from time to time to clean it; company literature suggested disassembling the magazine and

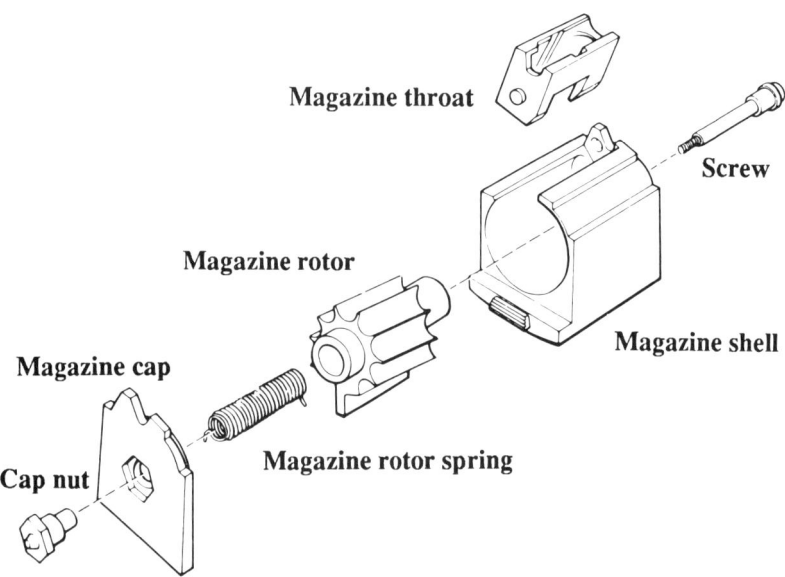

Exploded view of the 10/22 magazine assembly. (Drawing courtesy of Sturm, Ruger and Company)

The 10/22's magazine consists of seven different parts. The lips are made of steel for better wear-resistance. A coil spring, placed inside the plastic magazine rotor, rotates shells upward to be stripped off the feed lips of the magazine.

Variations

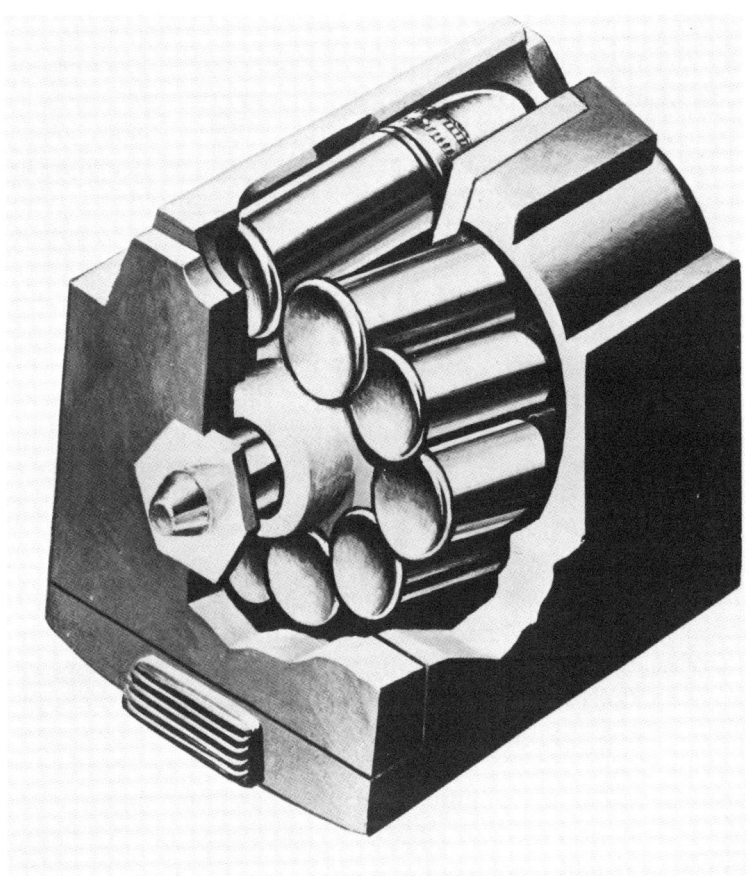

The cutaway drawing showing how 10 .22 rounds wrap around inside the magazine's rotary assembly. (Photo courtesy of Sturm, Ruger and Company)

cleaning it every 500 rounds. This proved to be beyond the skills of most shooters, however, and the retardation of the cycling of the rifle's action, coupled with a tight chamber, reduces fouling to a minimum. Consequently, owner's manuals for the 10/22 currently recommend that the user *not* disassemble the magazine. This seems to have proven quite satisfactory; most grime is readily removed with a rag or standard cleaning brushes without disassembling the magazine.

One small plus of the 10/22 is that its internal hammer can be lowered by pulling the bolt halfway back and squeezing the trigger. Many shooters like to lower the hammer before storing their rifles, and dry-firing a .22 firearm can, over time, peen the barrel so that rounds won't chamber properly. The Ruger design avoids this problem. It should be noted that such a practice is not called for; modern steel springs won't "set" over time and the extra compression caused by a cocked hammer shouldn't diminish the tension of a properly tempered spring even if compressed for many years. Care should, of course, be taken not to lower the hammer when a live round is in the chamber, since the slightest jar might then set it off.

When studying current, as well as older, Ruger catalogs, one glaring "hole" is quickly noticed: the lack of a stainless steel version of the 10/22. While company officials do not indicate that such a firearm is in the works, it would seem highly likely that one might be offered in the not-too-distant future. Sturm, Ruger and Company currently denotes its stainless steel versions of other guns with a "K" prefix on the serial number; it is probable that such a system would be used on the 10/22 as well.

Another variation might be created through the addition of a folding stock. If such versions should become available, chances are that 10/22s with folding stocks would have an "F" prefix added to their designations. It is also possible, but not likely, that scope mounts might be molded into the top of the 10/22 receiver and the guns sold with scope rings; in such a case, this version would probably have an "R" suffix to denote the variation. Whether any of these variations will be created, however, remains to be seen.

Another variation that could be created by Ruger in the near future is one of chambering. Certainly the .22 WMR would be a probable choice for such a rifle, though larger pistol cartridges, such as the .380 ACP, might be adapted to blowback use in a rifle patterned after the 10/22. Only time will tell.

That being said, here are the various versions of the 10/22 that have been made since the rifle was first introduced:

1) The 10/22RB, which is the standard carbine with a barrel ring at the front of its fore grip. The stock's composition has changed over the years; originally walnut, it is currently offered in walnut as the 10/22R, in a slightly less expensive "American hardwood" as the 10/22RB, or with a laminated stock (10/22RB-Z), which has a zebra-like, camouflaged look to it since the different layers of wood are stained tan, brown, and green before being laminated together and shaped. Checkering has never appeared on any of these guns and the butt has always had a Sharps-style curve to it. Early stocks had blued-steel buttplates; newer models have black plastic plates.

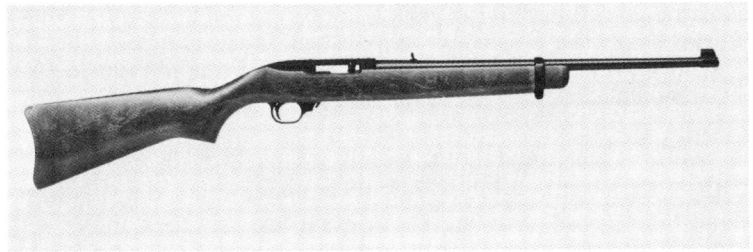

The 10/22RB is the standard carbine with a barrel ring at the front of its fore grip. Checkering has never appeared on any of the RB models, while the butt has always had a Sharps-style curve to it. Early stocks had blued-steel buttplates; newer models have black plastic plates. (Photo courtesy of Sturm, Ruger and Company)

The 10/22RB is available (as the 10/22RB-Z) with a laminated stock. The stock has a zebra-like camouflaged look to it, made with different layers of wood stained tan, brown, and green before being laminated together and shaped. (Photo courtesy of Sturm, Ruger and Company)

2) The discontinued Sporter, which lacks the barrel band of the original and has a fluted finger grip, a deeply curved grip, and a Monte Carlo comb (a raised ridge on the stock near the butt that improves use of a telescopic sight). Checkering adorns the Sporter's walnut stock at the grip and fore grip. Apparently Bill Ruger himself ordered the discontinuation of this stock shortly after it was introduced.

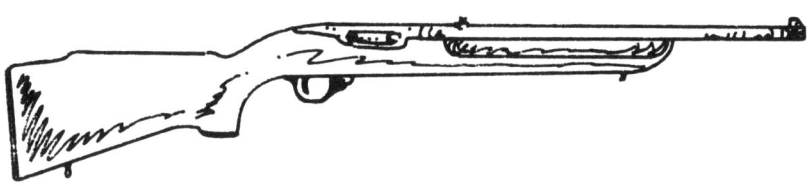

The discontinued 10/22 Sporter lacks the barrel band of the original and has a fluted finger grip, a deeply curved grip, and a Monte Carlo comb; the Sporter's walnut stock is checkered at the grip and fore grip. The same style stock was also used for the 10/22 "Centennial Gun" as well as for the Model .44 Carbine "Sporter."

3) The Centennial Gun, which was sold in a matched set with a Remington Model 742 to commemorate the Canadian Expo '67. These guns had the original Sporter stock above with a Monte Carlo comb with a medallion in the stock; serial numbers of the 10/22 and the 742 in each commemorative set were identical.

4) The Sporter SP (10/22SP), which replaced the original Sporter, had more classic lines, with a straight stock and rounded fore grip; the stock had sling swivels mounted on it.

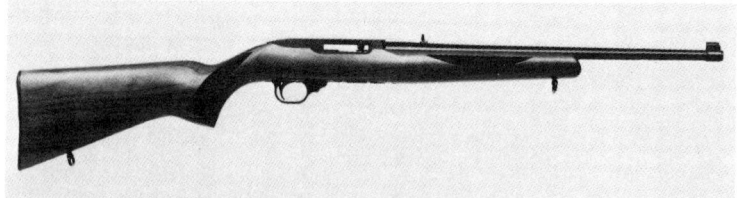

The Sporter SP lacks the barrel band of the standard model and has classic lines with sling swivels mounted on it. Checkering appears on the walnut stock at the grip and fore grip and the butt has less of a curve to it. Shown here is the deluxe version, the 10/22DSP. (Photo courtesy of Sturm, Ruger and Company)

Variations

Checkering was placed on the walnut stock at the grip and fore grip and the butt was straight. This is also available in a deluxe version as the 10/22DSP.

5) The International, which has a Mannlicher stock (a European stock that extends from the buttplate to the muzzle) that supports the barrel out to its end, where a barrel cap connects the barrel at its muzzle to the end of the stock. The International was discontinued because the stocks were hard to make; it was offered only from 1964 until 1970. Only a few thousand Internationals were ever made, which makes them more or less collector's pieces. Two versions of the International were offered; one was a plain, unchecked version, while the Deluxe International had checkering on the grip and fore grip.

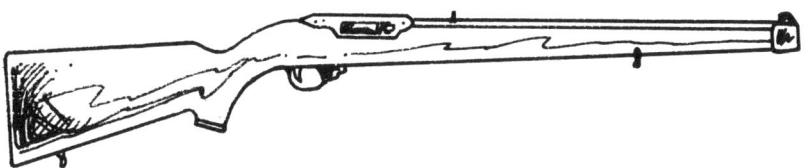

The discontinued 10/22 International has a Mannlicher stock that supports the barrel out to its end, where a barrel cap connects the barrel to the stock. The International's stocks were hard to make and only a few thousand of the guns were ever made. A similar stock was also offered for the .44 Carbine.

One notable external change to the rifle is in its front sight. The first 10/22s had an ivory insert like the front sight of the .44 Carbine. This was later changed to a small, gold bead, which has proven to be excellent in creating a small sight picture in all types of light conditions; such a sight helps milk the last bit of accuracy from the 10/22.

The 10/22 makes good use of its gold sight. Given good ammunition and a skilled shooter, worst-case shooting with a 10/22 will be four inches at 100 yards, since this is the performance requirement any given rifle must meet before passing final inspection at the Ruger factory. Most 10/22s actually exceed this by some margin. During a test by the American Rifle Association, the 10/22 grouped with 1.5 M.O.A. (minute of angle) with match ammunition and 3.48 M.O.A. with high-velocity rounds.

This translates to 1.5- to 3.5-inch groups at 100 yards. Most shooters discover that their rifles can fire 2- to 3-inch groups with some types of ammunition. The test also included a 1200-round functioning test; the 10/22 had no malfunctions during this extended period of firing!

10/22 Specifications

Barrel length	18.5 inches
Caliber	.22 LR
Length (approximate)	37.25 inches
Magazine (standard)	10 rounds
Rifling	6 grooves, right-hand twist
Weight (approximate)	5 pounds

THE MODEL .44 MAGNUM CARBINE

Since the 10/22 was made as a shooting companion to the .44 Magnum Carbine, Ruger wisely kept the basic layout of the .22 rifle nearly identical; while the .44 Carbine and the 10/22 operate quite differently, to a shooter interested only in the firearm's handling capability and outward appearance, the two carbines seem like variants of each other. From this standpoint, it is easy to think of the 10/22 as the .22 version of the .44 Carbine.

In addition, because the location of the safety, sights, operation, and so on are identical on both the 10/22 and the .44 Car-

The .44 Carbine served as the model for the basic layout of the 10/22 rifle. While the .44 Carbine is quite different from the 10/22 internally, when it comes to how the carbine is operated, the handling capabilities, and the .44's outward appearance, it is very similar to the 10/22. (Photo courtesy of Sturm, Ruger and Company)

bine, the rifles are ideal for shooters who may need to switch from one caliber to another. As many a shooter has discovered, having to fire different types of weapons during competition or hunting can be made easier if all the weapons operate in a similar fashion. Having different positions for safeties, charging levers, or magazine releases can cause an error on the part of the shooter. For hunters or others who use both a .22 rifle and a hunting rifle, the 10/22 and .44 Carbine make ideal companions.

While the .44 Carbine has been discontinued, many are still available on the market at fairly reasonable prices. It's also always possible that a new rifle, based on the .44 Carbine but using more investment castings, may one day be offered by Ruger; with the popularity of pistol-caliber carbines, it is possible that any new version of the rifle might be available in other chamberings in addition to .44 Magnum, with the 9mm Luger, .38 Special/.357 Magnum, and various .45 cartridges all being prime candidates.

Internally, the 10/22 and .44 Carbine are quite different. While the 10/22 operates on a blowback system, the .44 Carbine is gas-operated, using a short-stroke piston that is propelled backward slightly when gas, channeled through a small hole in the barrel, is forced against it when a round is fired. As the bullet leaves the barrel of the rifle, the gas pressure forces the bolt to unlock and the remaining energy of the piston propels the bolt backward, cocking the internal hammer and ejecting the empty brass cartridge in the process. As the bolt cycles back, the recoil spring finally overcomes its progress and forces the bolt forward to chamber another shell. At this point, the carbine is ready to fire again.

The tubular magazine of the .44 Carbine is located under the barrel. The shells in the magazine are pushed backward by a spring at the magazine's front and a shell lifter elevates them from the lower side of the receiver up to where the bolt can push them into the barrel's chamber. Shells are loaded into the magazine by pushing them against the lower side of the shell lifter so that they slide into the magazine, where they are locked into place and held until the bolt is cycled. The lifter latch is well shaped, so that cartridges are readily fed into the chamber by the bolt; a large extractor ensures that the large .44 brass is extracted efficiently.

As with the 10/22, there are several variations of the .44 Carbine:

1) The standard model, which has a barrel band at the front of the fore grip and lacks any checkering on its stock.

2) The M44SP sporter, which has the fluted, semi-beavertail fore end/Monte Carlo stock found on the first sporter version of the 10/22. Like its counterpart, this style stock was quickly pulled from the Ruger lineup.

3) The International, which has a Mannlicher stock that supports the barrel out to its muzzle. As with the 10/22 International, the .44 Carbine International was discontinued because the stocks were hard to manufacture. Both regular and deluxe grades were offered in the International model.

4) The 44RS, which was nearly identical to the standard model but had a built-in peep sight in its receiver and sling swivels.

5) The YR-25, which was a special commemorative version of the .44. The last of the .44s made, this rifle had a scene on it that depicted its twenty-five years of manufacture by Ruger.

As was the case with the 10/22, early models of the .44 Carbine have an ivory insert sight, while later models adopted the gold-bead sight pioneered on the 10/22; a built-in peep sight, located toward the rear of the receiver, was also offered. Upper receivers on the various models of the .44 Carbine are drilled and tapped for scope mounts.

Changes were also seen in the .44 Carbine's magazine-release system, which had a button added to the inside wall of the loading port so that cartridges could be released from the magazine without cycling the action.

While most shooters know that using pointed bullets in a tubular magazine can be dangerous, since the pointed bullet can ignite the primer of the shell in front of it during recoil, many are not aware that round-nosed or even nearly blunt bullets can have the same effect with the recoil of the .44 Magnum. Tests conducted by the Hornady Company with full metal jacket (FMJ) bullets with slightly rounded tops caused cartridges in the magazine of a Ruger .44 to ignite when a round was fired in the chamber, thereby demolishing the rifle. Therefore, care must be

The 44 RS model of the .44 Carbine was nearly identical to the standard model except for a built-in peep sight in its receiver and sling swivels mounted on the front barrel band and on the stock. Note the old-style ivory insert on the front sight. (Photo courtesy of Sturm, Ruger and Company)

The .44 Carbine YR-25 was a special commemorative version and the last of the .44s made after twenty-five years of production. The medallion in the stock makes this gun a collector's rifle rather than a "shooter." (Photo courtesy of Sturm, Ruger and Company)

taken when using the .44 Carbine to use only flat-nosed bullets in the tubular magazine. While this forces shooters to give up some potentially superior ballistics, it also avoids damage to the gun and a possible trip to the hospital.

The .44 Carbine is an excellent deer gun and a fair self-defense rifle, being limited in the latter use only by a rather small capacity magazine.

.44 Carbine Specifications

Barrel length. . . . . 18.5 inches
Caliber . . . . .44 Magnum
Length (approximate). . . . . 36.75 inches
Magazine . . . . 4 rounds
Rifling . . . . 12 grooves, 1-in-38, right-hand twist
Weight (approximate) . . . . 5.75 pounds

Chapter 3
Care & Maintenance

With a little care and know-how, the useful life of the 10/22 or .44 Carbine can be greatly extended; poor care can quickly spoil either gun or even destroy it in short order. With proper care, either of these fine Ruger firearms can easily last a lifetime.

CLEANING

The first order of business when cleaning a firearm or doing any other type of work on it is to be sure it's unloaded; many people have been injured or even killed when cleaning "empty" firearms. *Always* check to be sure a firearm is unloaded before casually handling it or doing any maintenance work.

Accuracy can be ruined by improper cleaning; in fact, it's better not to clean a rifle's bore at all rather than do a halfway job of it. The main consideration when cleaning a rifle is to clean the bore without damaging the muzzle end of the rifling. This means that the cleaning rod must not rub or bind against the bore at the muzzle. Ideally, rifle barrels are cleaned by pushing or pulling the cleaning brush and patches through the gun from the breech to the muzzle. This is not practical on the 10/22 (except with the Ram-Line Take-Down kit, which allows the barrel to be quickly removed) or the .44 Carbine; therefore, these guns are normally cleaned using a cleaning rod extended into the muzzle end.

Be especially careful with the cleaning rod and don't let it rub against the edge of the muzzle. This is best done by guiding the rod carefully into the barrel while using your fingers to keep the rod from rubbing against the muzzle. Take your time--going fast doesn't clean the gun any better than going slowly. When using patches, remember to push dirt in one direction only; going back and forth with a patch tends to just shove fouling up and down the bore without working it out of the gun. Shove the patch down the muzzle of the barrel, remove it through the chamber, and pull the empty cleaning rod carefully out of the barrel to repeat the process until the patches come through relatively clean.

Aluminum cleaning rods can become like files if they pick up grit; great care must be taken to keep an aluminum cleaning rod religiously clean. Perhaps a better choice is the U.S. military surplus steel cleaning rod; while steel can cause more damage to a bore than a clean aluminum rod, the fact that the harder metal won't pick up grit often makes it a better choice over the long haul. The standard cleaning kit for the M16 rifle will work for the 10/22, since the caliber of the .223/5.56 NATO round is close in diameter to the .22 LR. With the .44 Carbine, a cleaning brush made for the military's .45 automatic pistol will work fine, as will most commercial .44- or .45-caliber brass brushes.

Break-Free CLP is the lubricant/cleaner of choice for firearms. Created for U.S. military firearms, this lubricant can be used over a wide temperature range and is made to both clean and lubricate, so that bore cleaning fluid is no longer needed. Break-Free does such a good job of cleaning that Army manuals have had to be rewritten to tell inspectors to do away with white glove inspections since dirt continues to come free to be easily wiped off hours after the initial cleaning of a rifle!

Because Break-Free CLP takes time for its silicon lubricants to set up, let a carbine sit for a half hour or so after applying the fluid for full lubrication. Both Break-Free CLP and military surplus cleaning kits with steel rods are available from Sierra Supply.

The one place Break-Free CLP should not be used is in the

10/22's magazine. Since Break-Free CLP is a penetrating oil, it will quickly deactivate any .22 ammunition it chances to touch when the magazine is loaded. When cleaning the magazine, a soft brush and a powder fouling solvent are good bets. Do test a drop of the solvent on the outside of the magazine to be sure it won't attack the plastic.

Don't try to disassemble the 10/22's magazine unless something is wrong with it or it is impossible to remove dirt inside it; getting the proper tension on the spring is difficult once the spring has been removed. Be sure to check the slotted screw at the front of the magazine and, if it is loose, tighten it. If this screw comes out, the rotor shaft will lose its tension and will need to be readjusted (which is the most difficult task in reassembling the magazine).

Generally, the 10/22 should be cleaned after every 500 rounds, or several times a year if it isn't fired much. Because it is a hunting rifle, the .44 Carbine will probably need to be cleaned every three or four months, since not all that many rounds will probably be fired through it; cleaning a rifle that has not been fired can consist of simply running a patch soaked in Break-Free CLP down the bore to prevent it from rusting.

A *very* light coat of oil on steel parts will help prevent rust. When you've finished cleaning the rifle, check the bore to be sure it is clear of patch threads and oil. Any material left in the barrel can ruin it with just one shot.

Break-Free CLP or any other type of lubricant can be damaging to a firearm if used improperly. With the .44 Carbine, don't oil the area around the gas tube or the piston; oil in these areas can char and damage the assembly or cause poor functioning. With either the 10/22 or .44 Carbine, avoid getting oil on wooden stocks, as it can gradually stain the wood or even damage it. Care must also be taken not to leave oil in magazines, or in the chamber, where oil can deactivate ammunition or create excessive pressure, since it will keep the brass cartridge from sealing the chamber properly when fired.

Be careful not to use too much oil on internal parts, since this doesn't improve the lubricating properties of the oil, and may actually gum things up as dirt becomes trapped in the oil. Too much oil will actually increase the wear and tear on either rifle. There is also no need to lubricate the outside of the receiver of the 10/22 or its trigger guard, since these parts are aluminum.

If it's necessary to store either a 10/22 or .44 Carbine for several months or more, be sure the firearm is not placed in a leather or plastic container, which will collect moisture. Airtight containers will rust up a firearm very quickly, because moisture will condense inside them when the temperature varies even a few degrees.

Touch-up blue is useful for darkening down edges when the bluing has worn off or for filling in any nicks or scratches on the finish of the blued firearm. This is especially useful if you decide to sell a firearm, since the touch-up blue can do away with worn spots to make a rifle look as good outside as it is inside.

Minor rust spots can be removed from blued areas of a carbine and the surface refinished with touch-up blue. To remove the rust spot, lightly sand it with very fine steel wool soaked in oil; touch-up blue is then used according to the instructions on the bottle. Touch-up blue does not cover major scratches or large flat patches very well but does work well in touching up light wear and tear or rust spots. One important point to keep in mind is that touch-up blue often promotes rust if the area it is used on isn't cleaned off and lightly oiled after the chemical is used.

A black permanent marker can be used to touch up the aluminum receiver and trigger guard on the 10/22, but this is very temporary at best. A better bet for refinishing the aluminum parts of the 10/22 is to remove the barrel and bolt assembly and very carefully spray paint the areas showing wear. This is not an easy job, however, since great care has to be taken to strip the parts of oil (acetone is often used for this), match the original color of the finish, and keep paint out of areas where it will cause binding or gum up parts. Since a 10/22 functions just as well with wear showing on its aluminum parts, and since the aluminum can't rust or even corrode to any great extent, it is generally better to leave well enough alone and live with any minor wear on the finish.

Sheath anti-rust cloth, marketed by Birchwood Casey, works well and is ideal for use once or twice a year. Birchwood Casey also markets products to help clean and maintain wooden stocks. These products, as well as touch-up blue, are available at most gun stores.

Care & Maintenance

Linseed oil is quite useful for restoring the finish on a somewhat worn stock, as well as for touching up scratches. Linseed oil also helps waterproof a wooden stock somewhat; applying it on the inside of the stock as well as externally can really help keep the stock from swelling in humid or damp environments.

Neither the 10/22 or .44 Carbine require any special tools for maintenance and only a screwdriver is needed for fieldstripping. For detail stripping, a good drift punch or two, needle-nosed pliers, and a rawhide or rubber mallet are all quite useful.

Nearly as important as good tools is a soft surface on the workbench so that your firearm doesn't become scratched while you work on it. Small containers or an ice cube tray are essential for holding small parts during major disassembly.

Of major importance is not tackling more gunsmithing work than you can handle or taking a rifle apart that is working well. More firearms are damaged by improper disassembly or reassembly than by wear and tear. Leave the trigger group assembled when cleaning it; while the trigger group isn't overly complicated, several parts do double duty and are a bit hard to reassemble.

On the .44 Carbine, the recoil force tends to cause some parts to work loose; the cure for keeping the lifter latch pins, trigger pin, lifter cam pin, and two bolt pins in place is to stake them. This procedure is carried out by using a drift punch to peen a small amount of metal from the trigger housing or bolt against the ends of the pins to hold them in place. With care, this job can be done by the owner of the firearm, but when in doubt, let a gunsmith do it.

FIELDSTRIPPING THE 10/22 CARBINE

The various models of the 10/22 fieldstrip almost identically, except for differences in loosening the barrel band and receiver screw on the Carbine model as opposed to removing only the receiver screw on the Sporter. All models of the 10/22 are identical internally, with nearly all parts interchangeable.

On the rare International model, the tolerances are quite tight, especially on the end piece fastening the muzzle end of the barrel to the front of the stock. If this end piece fits tightly, removing it from the stock by loosening the small screw on its underside is often necessary.

To fieldstrip the standard 10/22:

1) Remove the magazine and cycle the weapon to be sure it's empty.

2) Unscrew the receiver/stock screw and the barrel band screw. Take the barrel band off over the muzzle of the barrel.

3) Place the safety in its middle position (on tight-fitting stocks), lift up on the barrel, and pull the barrel/receiver/trigger group up and forward out of the stock.

4) Drift out the two cross pins holding the trigger group in place and remove this assembly from the receiver.

5) Drift the bolt-stop pin out of the rear of the receiver.

6) Pull the bolt toward the rear of the receiver and lift it out through the bottom of the receiver.

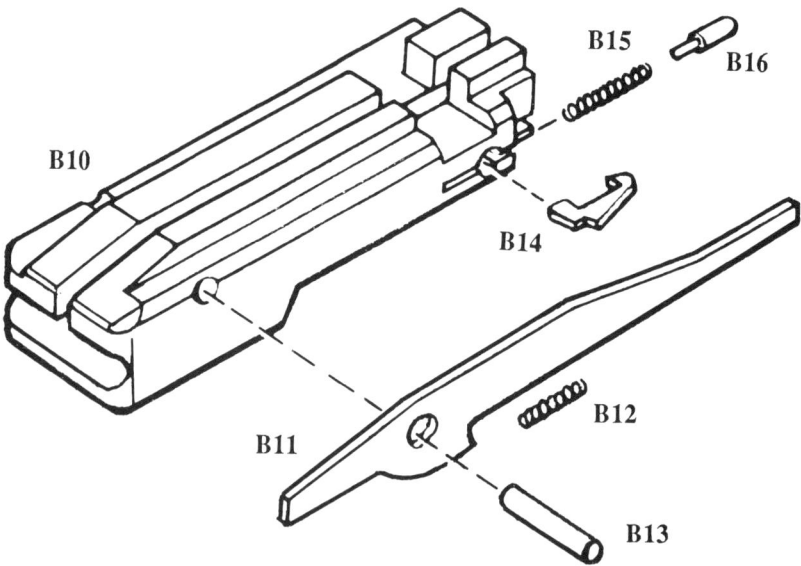

Exploded diagram of the 10/22 bolt assembly. (Drawing courtesy of Sturm, Ruger and Company)

B10 Bolt
B11 Firing pin
B12 Firing pin rebound spring
B13 Firing pin stop pin
B14 Extractor
B15 Extractor spring
B16 Extractor plunger

Care & Maintenance

7) Pull the charging handle, recoil spring, and guide through the side of the receiver. Do not disassemble this group unless necessary for repair or replacement of parts, since the small clip holding the recoil spring on its guide is rather hard to replace.

This is generally as far as disassembly is carried out. Normal maintenance and lubrication can be accomplished with only this much disassembly. If the barrel needs to be removed, this can be accomplished by removing the two bolts holding the barrel retainer in place and then wiggling the barrel out of the receiver. Most of the parts in the trigger group can be removed by drifting out the pins holding them in place; the hammer spring can be removed by placing the hammer in its forward position and then pulling the spring and its guide back and out of the trigger group. Avoid removing the safety if possible, since its spring is rather hard to replace; the safety can be removed by placing it halfway between the safe and fire positions and then rotating it a quarter turn and pushing it out.

Reassembly of the 10/22 is basically a reversal of the above procedure. One sticky spot is the bolt. Though getting the bolt into the receiver isn't much of a trick, getting it in properly is. The secret is to place the recoil spring on the rod-like protrusion at the left rear of the receiver and then reach into the underside of the receiver with a finger and pull the charging handle back with the finger. Once the charging handle is back and the recoil spring fully compressed, it is held back with the other hand by grasping the charging handle outside the receiver. With the charging handle secure, dropping the bolt back into the receiver (firing-pin side to the top of the receiver) should allow the bolt and charging handle to move forward without binding.

Replacing the bolt-stop pin, the trigger assembly, and the pins holding the trigger assembly in place gets you almost home. Be sure not to let any of the pins or the safety get caught on the stock when reassembling the rifle.

As noted elsewhere, the magazine of the 10/22 shouldn't be fooled with if possible, but it isn't quite as hard to reassemble as many think, either. If it should be necessary to disassemble it or if the screw holding it together pops out, all is not lost. The main trick is to keep the rotor spring inside the rotor, since removing it can create a major problem in reassembly.

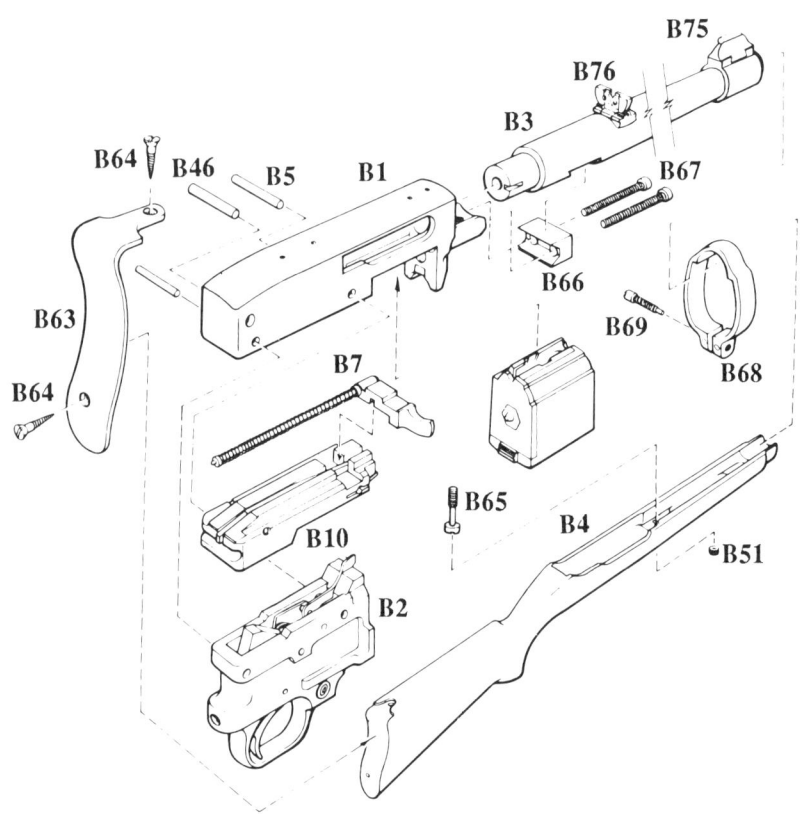

Exploded diagram of the 10/22 Carbine. (Drawing courtesy of Sturm, Ruger and Company)

- B1 Receiver
- B2 Trigger assembly
- B3 Barrel
- B4 Stock
- B5 Receiver, cross pin (2)
- B7 Charging handle, guide rod, and recoil spring assembly
- B10 Bolt assembly
- B46 Bolt stop pin
- B51 Takedown screw washer
- B63 Buttplate
- B64 Buttplate screw (2)
- B65 Takedown screw
- B66 Barrel retainer (V-block)
- B67 Barrel retainer screw (2)
- B68 Barrel band
- B69 Barrel band screw
- B75 Front sight
- B76 Rear sight

Care & Maintenance

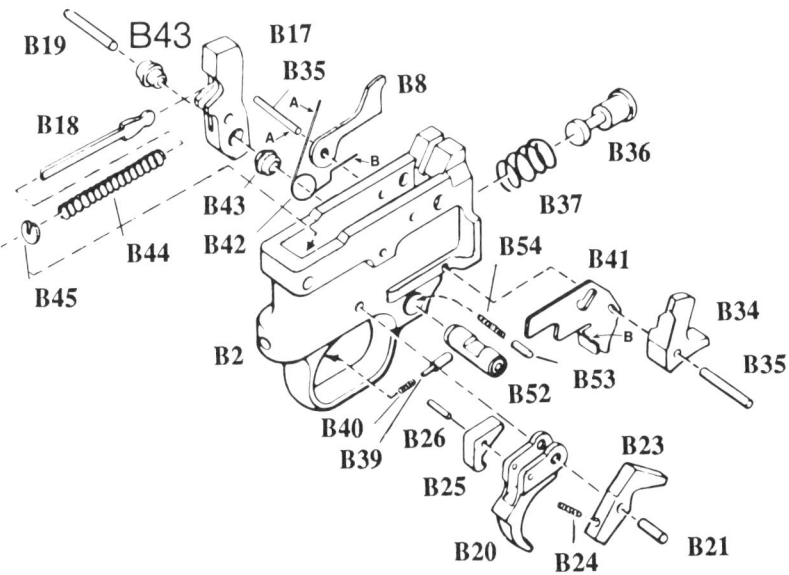

Exploded diagram of the 10/22 carbine trigger guard assembly. (Drawing courtesy of Sturm, Ruger and Company)

B2   Trigger guard
B8   Ejector
B17 Hammer
B18 Hammer strut
B19 Hammer pivot pin
B20 Trigger
B21 Sear and trigger pivot pin
B23 Sear
B24 Sear spring
B25 Disconnector
B26 Disconnector pivot pin
B34 Magazine latch
B35 Magazine latch pivot pin
      and ejector pivot pin (2)
B36 Magazine latch plunger
B37 Magazine latch plunger
      spring
B39 Trigger plunger
B40 Trigger plunger spring
B41 Bolt lock
B42 Bolt lock spring
B43 Hammer bushing (2)
B44 Hammer spring
B45 Hammer strut washer
B52 Safety
B53 Safety detent plunger
B54 Safety detent plunger
      spring

If the magazine is disassembled, the following procedure can be used to get it back together:

1) Place the long magazine screw into the front of the magazine.
2) Put the rotor and its spring over the magazine screw. The longer side of the hub of the rotor goes toward the front of the magazine (the end with the screw head).
3) Place the steel magazine throat on the top of the magazine, being careful that the bullet end is forward. The magazine throat should engage the larger hub of the rotor in the magazine throat's forward recess.
4) Place the magazine cap plate over the back of the magazine and hold the assembly together.
5) Place the hex nut onto the exposed rotor spring and rotate the hex nut until the spring's hook is retained in the hole in the hex nut. Do not push the hex nut into its recess in the cap plate face.
6) Turn the hex nut clockwise until the rotor can no longer rotate.
7) Once the rotor can no longer rotate, turn the hex nut another one-and-a-quarter turns to increase the rotor spring tension to its recommended adjustment.
8) Push the hex nut into its recess on the face of the magazine cap plate face and hold it in place.
9) Tighten the magazine screw while holding the hex nut in place. Take care not to overtighten the screw as it may cause the magazine to warp, binding the rotor inside it.

FIELDSTRIPPING THE .44 MAGNUM CARBINE

The various models of the .44 Carbine fieldstrip almost identically, except for freeing the barrel/receiver/trigger groups by loosening the barrel band on the Carbine model as opposed to removing the receiver screw on the Sporters. All models of the .44 Carbine are nearly identical internally, with most parts interchangeable.

As with the 10/22, the tolerances on the rare International .44 Carbine are quite tight, especially on the end piece fastening the

Care & Maintenance

Fieldstripped standard model .44 Carbine. (Photo courtesy of Sturm, Ruger and Company)

muzzle end of the barrel to the front of the stock. If this end piece fits too tightly, it is often necessary to remove it from the stock by loosening the small screw on its underside.

The following procedure should be used in fieldstripping the .44 Carbine:

1) Cycle the weapon to be sure it's empty, lock the bolt open, and check the magazine when the bolt is retracted to be sure it is empty.

2) Unscrew the receiver/stock screw or the barrel band screw and remove the barrel band over the muzzle of the barrel.

3) Place the safety into its middle position (on tight-fitting stocks), lift up on the barrel, and pull the barrel/receiver/trigger group up and forward out of the stock.

4) Close the bolt and drift out the cross pin holding the trigger group in place; pull this assembly down and back out of the receiver.

5) If necessary, you can take off the recoil block on the stock (at the rear of the receiver compartment) by taking off the butt-

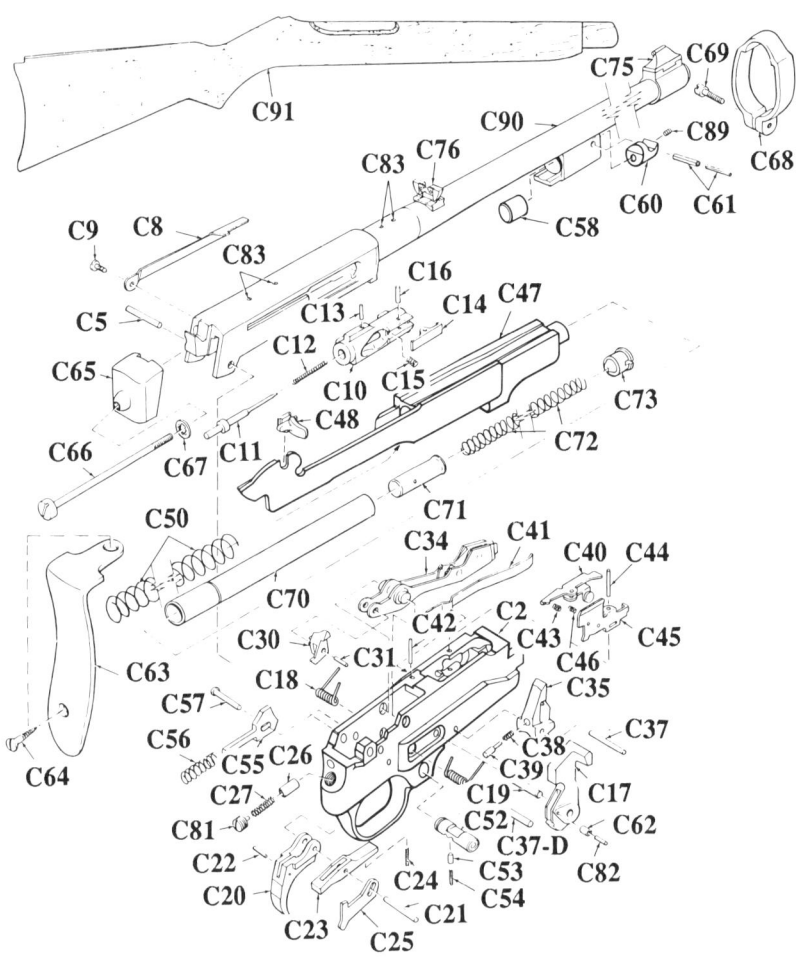

Exploded diagram of the .44 Carbine. (Drawing courtesy of Sturm, Ruger and Company)

Care & Maintenance

C2	Trigger guard (trigger assembly housing)
C5	Receiver cross pin
C8	Ejector
C9	Ejector screw
C10	Bolt
C11	Firing pin
C12	Firing pin spring
C13	Firing pin retaining pin
C14	Extractor
C15	Extractor spring
C16	Extractor retaining pin
C17	Hammer
C18	Hammer springs, left and right
C19	Hammer pivot pin
C20	Trigger
C21	Trigger pivot pin
C22	Trigger cross pin
C23	Sear
C24	Sear spring
C25	Disconnector
C26	Disconnector plunger
C27	Disconnector plunger spring
C30	Lifter dog
C31	Lifter dog pivot pin
C34	Lifter assembly
C35	Lifter latch
C37	Lifter latch pivot pin
C37-D	Hammer springs retaining pin
C38	Lifter latch plunger spring
C39	Lifter latch plunger
C40	Cartridge stop
C41	Cartridge stop flat spring
C42	Cartridge stop flat spring retaining pin
C43	Cartridge stop coil spring
C44	Cartridge stop pivot pin
C45	Flapper
C46	Flapper spring
C47	Slide
C48	Slide charging handle
C50	Slide spring
C52	Safety
C53	Safety detent plunger
C54	Safety detent plunger spring
C55	Lifter cam
C56	Lifter cam spring
C57	Lifter cam pivot pin
C58	Piston
C60	Piston block plug
C61	Piston block plug retaining pin (2)
C62	Hammer roller
C63	Buttplate
C64	Buttplate screw (2)
C65	Recoil block
C66	Recoil block pin
C67	Recoil block pin washer
C68	Barrel band
C69	Barrel band screw
C70	Magazine tube
C71	Magazine follower
C72	Magazine spring
C73	Magazine plug
C75	Front sight
C76	Rear sight assembly
C81	Disconnector plunger spring screw
C82	Hammer roller pivot pin
C83	Scope base screws/ screw holes
C89	Piston plug set screw
C90	Barrel
C91	Stock

plate and unscrewing the bolt that enters the rear of the block. Lift the block up and forward to remove it.

6) Hold the recoil spring and pull the action slide back and up slightly to free the recoil spring. Carefully snake the spring off the magazine and out the receiver.

7) Pull the charging handle out of its slot in the slide and remove it.

8) Pull the slot in the front of the magazine plug out of the rear of the piston block and remove the slide and magazine assembly from the barrel/receiver.

9) Remove the piston from the rear of the cylinder block.

10) If necessary, the magazine assembly can be disassembled by drifting out the roll pin at its front. Take care to restrain the magazine plug, since it is under pressure from the magazine spring. (Note the direction the magazine follower faces when disassembling the magazine assembly.)

11) To remove the bolt, first unscrew the ejector screw and remove the ejector, then rotate the bolt and push it to the rear so that its lugs clear the cuts in the receiver. This will free it so that it can be pulled out the bottom of the receiver.

This is generally as far as disassembly is carried out, though the firing pin and extractor may be removed for replacement by drifting out the two pins holding them in place on the bolt; take care not to lose the springs that are with these two parts. Further disassembly, especially of the trigger group, is rather tricky at best, and is a job better left to a gunsmith. Since normal maintenance and lubrication can be accomplished with only this much disassembly, the old saying, "If it's not broken, don't fix it," is especially appropriate with the .44 Carbine.

Reassembly is basically a reversal of the above procedure. When replacing the magazine tube, the larger half of the slotted end goes toward the barrel.

EMERGENCY REPAIR KITS

The Ruger .44 Carbine and 10/22 are often described as "gunsmith's nightmares," because they don't give this segment of the population much business. Because of this, emergency

Care & Maintenance

repair kits aren't really called for unless you're careless when cleaning, disassemble a gun and lose some parts, or monkey with the firearm and damage it.

Since .44 Carbine parts are not easy to come by and many 10/22 parts can be easily fabricated by a gunsmith, many owners may wish to just take special care of their guns and not order parts. For those who really worry about such things, however, a kit with a few essential parts that are prone to more wear than others might ease the mind. Parts to consider purchasing are extractors, ejectors, firing pins, recoil springs, and hammer springs, as well as some smaller parts that could be lost.

If you opt for purchasing parts, first try Ruger's Product Service Department, Newport, NH 03773, which sells parts for a nominal price; prices are usually listed in owner's manuals. Ruger guns that need to be repaired can also be shipped to this address. When ordering spare parts from Ruger, be sure to include your rifle's model and serial number, as there may be slight variations in parts from one model to another. Great care must be taken when replacing parts in any firearm since it is easy to ruin the gun's reliability if parts aren't properly assembled and/or fitted.

Because Ruger no longer makes .44 Carbines, it might be impossible to purchase certain parts from the company in the future. Should this be the case, companies specializing in selling parts from discontinued firearms should be checked after first contacting Sturm, Ruger and Company. One such source is Gun Parts (formerly Numrich Arms).

MODIFICATIONS TO THE 10/22 AND .44 CARBINE

Modifying any firearm should be approached with caution; poor modifications will make a firearm unreliable and even the simplest changes will usually void any warranty. Don't tamper with a firearm unless a change really needs to be made.

Probably the simplest modifications to make to these two Ruger carbines is to purchase "bolt on" accessories for them. Nearly all of these can be placed on the firearm by the owner, without the need for any gunsmith work. Because the .44 Carbine is out of production, accessories other than slings, scopes, and other items that are easily adapted to all types of rifles are hard to find. The 10/22, on the other hand, has a wealth of accessories--so many that they have to be covered in a later separate chapter.

With both guns, there are several modifications that might be considered. One consists of having a rear peep sight made to replace the factory notch sights. While the notch sight is easier to use initially for beginning shooters, it isn't nearly as quick as the peep sight and the view of the target is actually a bit sharper with a peep sight. These are both important considerations.

Some versions of the .44 Carbine were made with a receiver-mounted, adjustable peep sight; this would be first choice for the .44 Carbine, if such a gun can be found. Second choice for the .44 Carbine, and the only choice for the 10/22, is to replace the small notch blade in the rear sight with a new blade having a small aperture drilled in it or to purchase an aftermarket peep-sight that mounts on the carbine's scope mount (these aftermarket sights are covered in the next chapter).

It is hoped that some aftermarket manufacturer will offer such a blade in the near future; in the meantime, it's a job for a gunsmith or a skilled do-it-yourselfer who's handy with a drill press and file. Black spray paint or touch-up blue will get the new peep sight to the proper color to match the firearm.

Another change that might appeal to some shooters is to replace the front safety with a blade safety similar to that of the Garand/M14/Mini-14. This type of safety allows the rifle to be

One alteration of the 10/22 that aids in quickly sighting a target is the replacement of the carbine's notch blade in the rear sight with a new blade having a small aperture drilled in it.

placed in the fire position with the finger inside the trigger guard. While such a safety for the 10/22 or .44 Carbine would not be identical to that used on the Garand series of guns, operation would be similar enough to give carryover skills for a person switching from the 10/22 or .44 Carbine to one of the larger rifles. However, making such a safety is, again, a job for a gunsmith.

Most shooters are happy with the pull of their carbines but, for those who aren't, the pull can be lightened. Doing trigger work is very tricky at best; always leave this work to a gunsmith.

For those who use the 10/22 and don't own a .44 Carbine, a .22 that's nearly as heavy as a centerfire rifle becomes a doubtful asset. In such a case, a few things can be done to lighten the 10/22.

One is to purchase the Ram-Line 10/22 plastic barrel described in the next chapter. This will shave almost a pound from the rifle. Despite what one might think, changing to a plastic stock won't lighten the rifle; wooden stocks are generally as light as aftermarket stocks. It is possible to lighten wooden stocks a bit, however, by drilling out the wood beneath the buttplate (being careful not to cut through to the outside surface). This will shave a few ounces from the rifle's weight, though not as much as one might think.

For those not worried about the looks of the 10/22, a bit more weight can be taken from the gun by cutting off the stock just ahead of the barrel-retaining V-block. Refinishing the stock will make this new cut in the front look good, though, certainly, no one is going to mistake it for the original lines of the carbine.

For anyone who wants an AR-15-style pistol grip on his carbine, the easiest solution is to purchase one of the aftermarket plastic stocks. This isn't an option for those who own a .44 Carbine, however, and some 10/22 owners may balk at shelling out forty bucks for such a stock after buying a good stock with the rifle in the first place. For these people, it is possible to modify the factory stock slightly and add a pistol grip to it.

A few woodworking tools, some time, a little skill, and patience allow the job to be done *provided* the owner doesn't mind cutting into the stock that came with the carbine. In such a case, all that needs to be purchased is a military surplus M16/AR-15 pistol grip for about $6 from places like Sherwood International

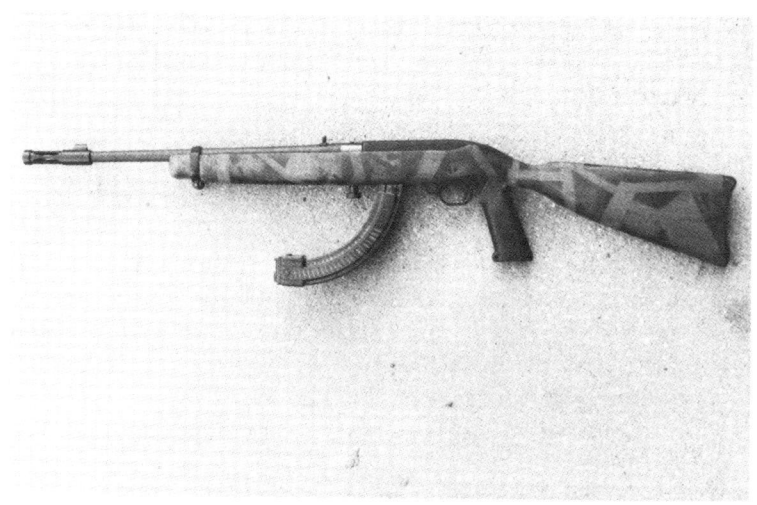

A little work with some woodworking tools plus the purchase of a military surplus M16 grip and some paint resulted in the cheap transformation of a standard 10/22 stock to this pistol-grip camouflage stock. The rifle also sports Ram-Line's flash hider, Ultralight barrel, Take-Down kit, and two 50-round magazines glued together.

or SGW. This done, a small portion of the grip on the lower side of the wooden stock needs to be removed to give thumb clearance, a small indentation the size of the top of the AR-15 grip chiseled into the lower side of the stock, and the grip screwed into place. Again, the stock will need to be refinished to look good. For the military look that seems to be popular these days, the finish can be stripped or lightly sanded and a flat black paint added to the stock. A little do-it-yourself designer work with hunter's camouflage tape or a similar paint job may also be to your liking.

In general, the fewer changes made in a firearm, the less something is apt to go wrong with it. Therefore, give a lot of thought to having a 10/22 or .44 Carbine modified.

TROUBLESHOOTING THE 10/22 AND .44 CARBINE

One of the main causes of failure in the Ruger carbines is damage caused when the firearms are improperly disassembled or reassembled, or modified by those who don't know what they're doing. Other failures are often caused by using poor ammunition or by extensive fouling (which is especially possible with extended shooting of cheap ammunition).

Some cartridges don't work well in the 10/22 and .44 Carbine. In the 10/22, cartridges with bullets with blunt noses or sharp shoulders will often cause jams. Explosive bullet loads can go off during chambering and should never be used in the 10/22. As noted elsewhere, care should be taken to use only bullets with flat or hollow points in the .44 Carbine so that recoil does not cause bullets in the magazine to fire the cartridges ahead of them. Reloaders should avoid trying to load the ultimate "hot" round as well; if the .44 Magnum isn't powerful enough in standard loads, it's time to switch to a rifle chambered for a heavier rifle round. Using lighter loads rather than hot loads will greatly extend the life of a .44 Carbine.

The Ruger guns will generally function at peak reliability after being broken in, because use causes rough edges to be rounded off and surfaces to be polished. After several hundred rounds, a firearm will often "custom fit" its parts and function very reliably from then on. Much the same type of fitting can be done by hand-cycling the weapon a number of times, but not all parts are affected, nor does this work nearly as well as actually shooting the firearm.

Avoid dry-firing any gun, since this can weaken or even break the firing pin over extended periods; with .22 guns such as the 10/22, the chamber may also become peened when the firing pin hits it.

Keep your carbine clean; while the Ruger carbines may continue to work well when extremely dirty, all weapons will fail eventually, given enough dirt in the action and chambers. In addition, grit or sand will cause excessive wear on a rifle so that its parts may wear out very quickly. A clean weapon lasts longer and is less apt to fail than an identical, but dirty, firearm.

At the other extreme of things to avoid is firing a round with a barrel obstruction. Oil, grease, a bit of cleaning rag, a misfired bullet, or a spider looking for a home in the barrel may cause the barrel to rupture or "goose egg" when the gun is fired. Always be sure the bore is clear and be careful to not get a plug in the barrel when in muddy areas or when it's raining or snowing.

When placing accessories on the 10/22 or making modifications to the rifle, it is wise to put on one accessory at a time or make one modification and then test the rifle out. If the rifle fails after piling on several accessories or making a number of changes, it's hard to determine the culprit. And don't purchase accessories for your rifle unless you really need them.

If a 10/22 or .44 Carbine fails to fire, there are some quick steps that you should go through to be sure that some simple thing hasn't caused the failure:

1) With the 10/22, tap the magazine to be sure it's seated.

2) Pull back on the charging lever again and check that a case is ejected and a shell isn't jammed in the weapon. If the chamber is clear, release the charging handle to chamber a new round. Do not "ride" the bolt forward; rather, let the recoil spring slam the bolt home with full force. To check to see if a round is chambered, pull back slightly on the bolt and peek at the chamber. Be sure to push the bolt fully forward after you've determined whether the chamber has a round in it.

3) Check the safety to be sure the rifle is ready to fire.

4) Try to fire again.

5) If the rifle fails to fire, go through steps one through four one more time.

6) If your carbine still fails to function, remove (on the 10/22) or empty (on the .44 Carbine) the magazine, cycle the weapon to be sure it's empty, and check out the inside of the ejection port while you pull back the bolt. This may reveal a problem in the form of dirt or a broken part.

7) If you haven't found the fault, change magazines (on the 10/22) or refill the magazine (on the .44 Carbine), cycle the action, and try firing again.

Even if you go through these steps and your rifle is clean and well maintained, it may still fail upon occasion. The best bet is simply to take the rifle to a gunsmith to get it repaired. If, however, you're in the middle of nowhere on a hunt or using the rifle in a combat or survival situation, then knowing how to get your carbine functioning again may be essential and in extreme situations might even mean the difference between life and death. In such a case, you should be familiar with the following procedures so that you know what to do if you should have to try to get your firearm into firing condition without the aid of a skilled gunsmith.

But be forewarned: *Many of the following procedures are dangerous.* Do not carry out any of these steps unless it is truly necessary. Generally, most cases in which a firearm fails will not warrant taking such a risk; take your firearm to a gunsmith.

It should be noted also that there are slight differences in some parts between the 10/22 and .44 Carbine; some steps may not apply to the firearm with which you're having problems.

10/22 AND .44 CARBINE TROUBLESHOOTING PROCEDURES

Problem	Check For	Procedure
Bolt does not hold open after last round (.44 Carbine)	Fouled/broken bolt latch	Clean/replace
Bolt won't lock	Fouling in locking lugs	Clean and lubricate lugs
	Recoil spring is not moving freely	Remove, clean, and lubricate
	Bolt is not moving freely	Remove, clean, and lubricate
	Gas piston misaligned	Check alignment; realign or replace
	Loose or damaged piston	Tighten or replace

Problem	Cause	Remedy
Bolt won't unlock (.44 Carbine)	Dirty piston	Remove and clean
	Dirty or burred bolt	Clean or replace
Firearm won't cock; safety doesn't work properly	Worn, broken, or missing parts	Check parts, replace
Firearm continues to fire after release of trigger	Dirt in trigger/sear	Clean mechanism
	Broken sear/trigger	Replace
	Weak sear/trigger spring	Replace
	Broken disconnector	Replace
Firearm won't fire	Safety in "Safe" position	Place in "Fire" position

10/22 AND .44 CARBINE TROUBLESHOOTING PROCEDURES

Problem	Check For	Procedure
	Firing pin is broken	Replace
	Too much oil or dirt in firing pin recess	Wipe/clean
	Poor ammo	Remove/discard
	Weak or broken hammer or hammer spring	Replace
	Bolt isn't locking (.44 Carbine)	Clean dirty parts
Round won't chamber	Dirty or corroded ammo	Clean or replace ammo

	Damaged ammo	Replace
	Fouling in chamber	Clean with chamber brush
Rounds won't eject	Broken ejector	Replace
Rounds won't extract	Broken extractor or bad extractor spring	Replace
	Dirty/corroded ammo	Remove (may have to be carefully pushed out with cleaning rod. Take care: *This procedure can be very dangerous.*
	Carbon/fouling in chamber or extractor lip	Clean chamber and lip
	Dirty/faulty recoil spring	Clean/replace

10/22 AND .44 CARBINE TROUBLESHOOTING PROCEDURES

Problem	Check For	Procedure
Rounds won't feed	Dirty or corroded ammo	Clean ammo
	Low-powered ammo or poorly shaped nose on bullets in cartridges	Use different ammo
	Defective magazine (10/22)	Replace magazine
	Dirt in magazine	Clean and lubricate magazine
	Insufficient gas to cycle action fully (.44 Carbine only)	Clean gas port, gas rods, etc.

Magazine not seated (10/22)	Reseat/replace magazine
Broken magazine catch (10/22)	Repair/replace
Broken shell lifter (.44 Carbine)	Repair/replace
Dirt/lack of lubrication	Lubricate; if it still binds, clean with soft brush
Poor ammunition	Replace
Fouling in gas port (.44 Carbine)	Clean gas port
Damaged piston or slide (.44 Carbine)	Repair/replace

Safety binds

Short recoil (new rounds fail to chamber)

Chapter 4

Accessories

There seems to be no end to the gadgets and accessories made for the 10/22. The .44 Carbine is a different story, since the rifle is out of production; while a few accessories like scopes and slings are available for it, there are no real aftermarket products made for this carbine. This may be just as well, since adding accessories can quickly weigh down a hunting rifle so that it's very tiring and awkward to carry. Because the whole point of using a .44 Carbine for hunting is that it's light and handy, turning it into an anchor is a rather foolish thing to do.

It can also be a foolish thing to do to the 10/22. This handy little carbine isn't an assault rifle or a military gun. Those needing such a gun should go with a real combat rifle like the Mini-14 or AR-15 rather than try to make the 10/22 into something it isn't by piling on military-looking accessories.

All that's really needed with either the 10/22 or the .44 Carbine is quality ammunition; not buying any accessories other than perhaps a sling and a good scope is a good idea for most people, while a few may benefit from a plastic stock or a lightweight plastic barrel, a flash hider (to keep from damaging the muzzle of the barrel), or perhaps some other odds and ends. But keep in mind that anything added to a carbine that just increases its weight is money poorly spent.

At the tail end of a list of accessories would have to be

bayonets, fast-fire devices, and barrel shrouds. While some argue that such devices can be intimidating if the rifle must be employed for self-defense, such accessories also make the firearm more awkward to use quickly.

That having been said, here's a look at some of the gear and gadgets available to owners of these two Ruger carbines.

BARRELS

There's nothing wrong with Ruger's barrel on the 10/22, except that, being steel, it does add some weight to the firearm. A bit of this weight can be taken off by having a gunsmith lop a few inches off the muzzle end of the barrel. Because the barrel is easily removed, however, an easier route (and one that is actu-

Ram-Line's high-tech Ultralight barrel is made of plastic, with a steel liner (containing the rifling) and a steel chamber. The mostly plastic barrel weighs only half as much as a steel barrel, which means that replacing the standard barrel on a 10/22 with an Ultralight results in a rifle that's a full pound lighter. (Photo courtesy of Ram-Line)

Accessories

ally a bit less expensive, since the front sight of the rifle doesn't have to be rebuilt) is to purchase Ram-Line's new high-tech Ultralight 16-inch barrel.

The Ultralight barrel is made of composite plastic, with a steel liner that contains the rifling, and a steel chamber. While more than strong enough to withstand the pressures created by the hottest of .22 LR rounds, this barrel weighs only half as much as a steel barrel. This means that replacing the barrel on a 10/22 with an Ultralight results in a rifle that's a full pound lighter. In addition, the carbine's center of balance is moved farther toward the receiver, making it seem even lighter. For those wanting a light .22 rifle coupled with the high reliability of the 10/22, this barrel is the way to go.

The Ultralight barrel comes with a front sight; the rear sight of the Ruger barrel is easily drifted out to be replaced in the plastic barrel, which has a cut in it to receive the sight. It is also possible to drift out the front sight of both the Ruger barrel and the Ultralight and place Ruger's gold-ball front sight on the plastic barrel; some may prefer this to the glow-orange sight that comes with the Ultralight.

The muzzle of the Ultralight is crowned so that the rifling is slightly recessed and protected from damage through bumps or scrapes. Cost of the Ultralight barrel is about $39.

BARREL SHROUD

E & L Manufacturing makes an aluminum ventilated shroud that goes over the 10/22 barrel and is fastened in place by a forward grip similar to that used on the HK-94 barrel shroud. Frankly, it's hard to imagine any time when this really would be necessary, since .22s don't generate much heat and a shooter's hand would rarely be on the forward end of the barrel in any case. For those wanting to dress up their 10/22s, the E & L shrouds cost about $35.

BAYONETS

A bayonet on either the 10/22 or .44 Carbine is of only slightly less use than would be an electric can opener. The spike bayonet--the only shape that really works in combat--has long since been abandoned for bayonets that can be used as combat knives when not mounted on a rifle. A case in point is the new U.S. Army bayonet, which is actually a fighting blade that can be mounted on the rifle for parades or for scaring crowds during demonstrations; while older Army bayonets had narrow blades that allowed them to be marginally effective, the new bayonet's blade is much too wide; friction would make it nearly impossible to pull out of an opponent's body.

At any rate, there aren't, at the time of this writing, any commercial bayonet adapters for the 10/22 or .44 Carbine. Still, such adapters have enjoyed brisk sales for shotguns and sporters such as the Mini-14; it's probably only a matter of time before such an adapter becomes available. Don't yield to temptation and waste money on such a device unless you're concerned about running out of ammunition when being attacked by marauding rabbits.

BIPODS

Bipods are useful for target shooting or displaying a rifle, but out in the wilds, most bipods are too short to allow you to see what you're shooting at and they're a little slow to get into play. Unless you're using your carbine for varminting or target shooting, you'll do well to forget a bipod.

Currently there are several types of "clothespin" bipods that clip onto the barrel of a rifle. The military bipod designed for the AR-15 doesn't work well, since it's made for a larger diameter barrel than that found on the 10/22 or .44 Carbine. It will work *if* you glue some rubber strips inside the bipod's mouth or wrap material around the rifle's or carbine's barrel.

A much superior and lighter clothespin bipod is Ram-Line's nylon model, which will fit almost any diameter barrel. This plastic bipod is also less apt to mar a carbine's finish. The Ram-Line bipods cost about $15 and can be quickly mounted or removed; carrying them is easy, too, since the bipod legs lock together so the device can fit into a large pocket or carrying case.

For those needing a high-quality, adjustable bipod, the Harris bipod is first choice. While the Harris Ultralight bipods look complex, they're simple to use and can be mounted onto a detachable front sling swivel base. Uncle Mike's swivels work perfectly for this and can be purchased at most gun stores for a few dollars. Because the Harris bipod attaches to the sling swivel mount, there is absolutely no problem with barrel flex; the rifle's zero stays the same if the bipod is in use or if you're standing and firing with the bipod still attached. Two models are currently offered: an 8- to 13-inch adjustable bipod, which is ideal for prone shooting; and a 13- to 23-inch bipod that allows shooting from a seated position and is ideal in the real world where a lot of tall weeds and grass always seem to get between the target and shooter, making prone shooting impractical. Both versions sell for about $40 each.

CARRYING CASES

To protect a carbine from getting scuffed and nicked, a carrying case is a must. Cases are also ideal for storing firearms in a closet or car trunk for long periods of time *if* air can circulate through them so that moisture doesn't condense inside the case and rust the gun. A case is also ideal for keeping a low profile when carrying the rifle through a populated area.

A hard plastic or aluminum case is a must if you're transporting a firearm on a commercial airplane (be sure to check on regulations and do *not* let a clerk put a tag on the outside of the rifle case if at all possible--cases so marked are often stolen). A hard case is also a good idea if you are taking a four-wheel drive vehicle through backcountry.

Hard cases cost big bucks. If you only need a carrying case occasionally, it makes sense to get one of the less expensive cloth cases. Best are those made of heavy canvas, which generally are available in camouflage, olive green, brown, or black. Parellex, Sherwood International, and a number of other companies carry these for under $30.

For carrying a rifle cross-country on a horse, snowmobile, or motorcycle (again, check local laws before you do this), a nylon scabbard from Uncle Mike's is ideal. The nylon scabbard can double as a carrying and storage case so that one purchase can set you up for all-around use. Cost for a carbine-sized scabbard is about $45.

FAST-FIRE DEVICES

The short trigger pull of the 10/22 makes it possible to fire the carbine very rapidly. Many shooters like the idea of being able to fire the weapon in a "rock and roll" automatic mode. Unfortunately, with current legal restrictions on automatic firearms in the United States, selective-fire 10/22s are about as common as snail toenails. Law enforcement and government organizations aren't restricted from creating automatic weapons, however. One excellent "how-to" book that gives the details on how to convert a 10/22 into a selective-fire weapon without modifying the trigger or receiver housing is *The Ruger 1022 Exotic Weapons System,* available from Paladin Press for about $15.

Several devices have been marketed to fill the demand for fast-fire weapons. These get around the legal restrictions against automatic weapons by enabling the shooter to turn some sort of crank or ratchet, which depresses the 10/22's trigger more quickly than can be done manually. Since the 10/22 still only fires with one pull of the trigger, it is only a semiauto weapon and not subject to the restrictions on selective-fire guns.

Perhaps the best known of these devices is the BMF Activator. This device is fastened onto the trigger guard with two knurled knobs. Turning the crank causes a small tongue to hit the trigger several times for each turn of the handle. High rates of fire can be achieved with the BMF Activator, and most people playing around with the device will probably want to couple it with a 50-round Ram-Line magazine.

Unfortunately, the plastic of the BMF Activator warps if left on a firearm so that it is necessary to remove it after it's been used. Shooters must take care to leave the 10/22 unloaded and the safety on until the moment of firing with the BMF Activator, since the handle is exposed and accidentally turning it could cause a string of shots to be fired.

Another variation on the fast-fire theme that, again, can be used without a special permit is the Ultimate, which fires at rates of up to 450 to 600 rounds per minute. The Ultimate consists of a pistol grip that is fastened to the standard stock of a 10/22 by a clamp similar to that used for plastic pipes. The grip has a curved "trigger guard" that allows the shooter's whole hand to fit inside it. A lever is then worked with the first fingers of the

Accessories 63

hand; when the lever is pulled back, a shot is fired and, when the lever is pushed forward, the device again depresses the trigger to fire a second shot. This doubles the rate of semiauto fire a shooter can achieve. The Ultimate is available from Firearm Systems and Design for a little more than $100, with models available for the AR-15 and M1 Carbine as well. The company also offers a forward grip that is mounted on the stock (à la Thompson submachine gun) with two clamps.

Perhaps the most outlandish of the fast-fire accessories is the Calico "Two-Twenty-Two" which, when assembled, looks like a miniature twin-barreled antiaircraft gun. This accessory consists of a tripod and a gun mount that accepts two stockless 10/22 rifles each placed into the contraption with their magazine wells pointing outward toward the left and right. A front sight/flash hider holds the barrels of the two rifles together at the front while the kit's barrel shroud covers the barrels.

The Two-Twenty-Two has an adjustable sight mounted on its main mechanism, a large, knobbed flywheel on its right side, and a pistol grip with the trigger for the dual-barreled firearm at its rear. When the unit is set up and the two rifles loaded and

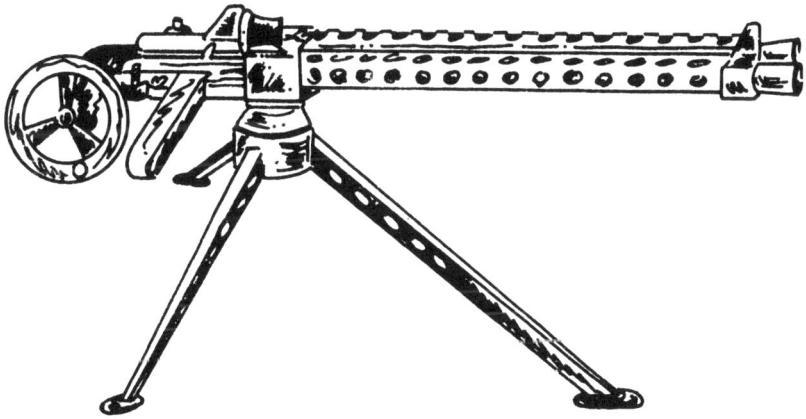

The Calico Two-Twenty-Two is a fast-fire accessory that, when assembled, looks like a miniature twin-barreled antiaircraft gun. When the unit is set up, cranking the wheel while pulling back on the trigger fires each of the two 10/22 rifles in the unit two times for a total of four shots per rotation. By cranking the unit quickly, very fast rates of fire are possible.

cocked with their safeties off, cranking the wheel while pulling back on the trigger fires the contraption. As the wheel is turned, small cams inside each of the two rifles' trigger guards push the triggers so that a very fast rate of fire is easily achieved. One turn fires each rifle two times for a total of four shots per crank. The kit comes with all necessary parts except the two 10/22 rifles, as well as the hex wrenches needed for assembly. All parts are tough and made of either aluminum castings or sheet steel. Adjustment of the Two-Twenty-Two can be a little tedious but, once working properly, the assembly burns ammunition so fast the pocketbook can start to smoke.

Accuracy is not great, because the point of aim tends to wander as the flywheel is cranked. The Two-Twenty-Two *looks* mean, but is really more toy than weapon (albeit a dangerous toy if not used properly). However, the gizmo is a lot of fun for those who get a kick out of plinking. Calico makes similar versions of the Two-Twenty-Two for the Mini-14 and the M1 Carbine; the kits are distributed by Mitchell Arms and cost about $290.

FLASHLIGHT MOUNT

Choate Machine and Tool offers a steel flashlight mount designed to clamp a mini-flashlight using AA batteries to the barrel of a 10/22. For the brave soul who needs to use a .22 rifle to defend himself against a burglar or for some types of hunting this unit might be a good buy.

The mount consists of two steel plates connected by bolts. Two channels run between the plate faces, so that when the plates are screwed together, one channel clamps onto the 10/22's barrel and the other to the flashlight. The plates have a one-and-a-quarter degree angle between them, which tilts the flashlight up ever so slightly to place the beam where the bullet will be hitting at closer ranges.

Care should be taken when using such a device in the dark, since it is often hard to see clearly, especially the area outside the beam's radius, and because anyone lurking in the dark can see exactly where you are. If you're using the rifle for self-defense, bear in mind that the flashlight, with you behind it, makes a nifty target.

Accessories

Choate Machine and Tool's steel flashlight mount is designed to clamp a mini flashlight to the barrel of a 10/22 (shown here with Choate's fixed stock). The unit consists of two steel plates connected by bolts with a one-and-a-quarter degree angle between the plates. This causes the flashlight's beam to hit at the same point at which a bullet fired from the rifle at close ranges will strike. (Photo courtesy of Choate Machine and Tool)

FLASH SUPPRESSORS AND MUZZLE BRAKES

At first glance, a flash hider on a .22 rifle looks like a terrible weekend warrior type of joke. But in fact, such an item does have a good use: it helps keep dirt out of the barrel and keeps the muzzle from getting dinged up. For those who may be using the 10/22 under adverse conditions, placing a flash suppressor over the muzzle of the barrel is a wise move.

Several companies offer flash hiders for the 10/22. Two of the best are offered by Ram-Line, which makes a light aluminum model for about $15, and Choate Machine and Tool, which offers a small blued-steel model, also about $15. Each slips over the barrel and is held in place by a hex screw; the front sight can then be used and doesn't need to be replaced.

The Ram-Line flash hider is a full-size unit with the look of a unit designed for a .223 assault rifle, while the smaller Choate flash suppressor adds dog ears similar to those on a military rifle to either side of the sight. Because Choate's steel flash hider is small, it weighs the same as the Ram-Line aluminum unit.

Another type of flash hider is offered by Mitchell. This is another slip-on flash hider, and is made of blued steel. Like the others, the Mitchell hiders are held in place with set screws. The standard Mitchell flash hider is good; another model, however, with a "muzzle brake," is of dubious use given the low recoil of the .22 LR. Cost for the standard model is about $20, while the muzzle brake model sells for about $23.

Finally, in the military surplus market, there is a rather crude conical, clamp-on flash hider made for the M1 Carbine. This device uses a bushing made of lead so that it can fit the various sizes of M1 Carbine. This bushing also makes it possible for the clamp-on flash hider to fit the muzzle of the 10/22. While not as pretty as the others, the unit may be cheaper if you look around a bit and want a flash hider that can be quickly removed. Sherwood International usually has these in stock.

MAGAZINES

The standard 10/22 rotary magazine is hard to beat. However, for those wishing for a bit more firepower when plinking, or those who might be forced to use a .22 for self-defense, a magazine holding more shells is often called for.

The catch with many of these aftermarket magazines is that they may not line up exactly with the feeding point of the 10/22. If this is the case, rounds will often jam in the rifle rather than chambering properly. Throating the barrel will usually solve this problem and bring the reliability of the rifle up to par. Throating consists of removing a small amount of barrel material at the base of the chamber where rounds feed into it, as well as rounding off the area all the way around the chamber opening. This creates a larger area where the nose of a shell can strike and still be pushed into the chamber during the firing/reloading cycle.

Such work should be carried out by a gunsmith if at all possible. While a do-it-yourselfer can modify a 10/22 easily, taking out too much metal will make the rifle dangerous because the brass cases don't have enough support, may make the rifle fail to fire, or might actually increase the number of jams. So the best bet is to take a quick trip to a local gunsmith when a throating job is called for.

Currently, Ram-Line makes the best extended "banana" magazines. The company carries 30- and 50-round models, each of which is quite reliable and has a special pair of coiled springs

Accessories

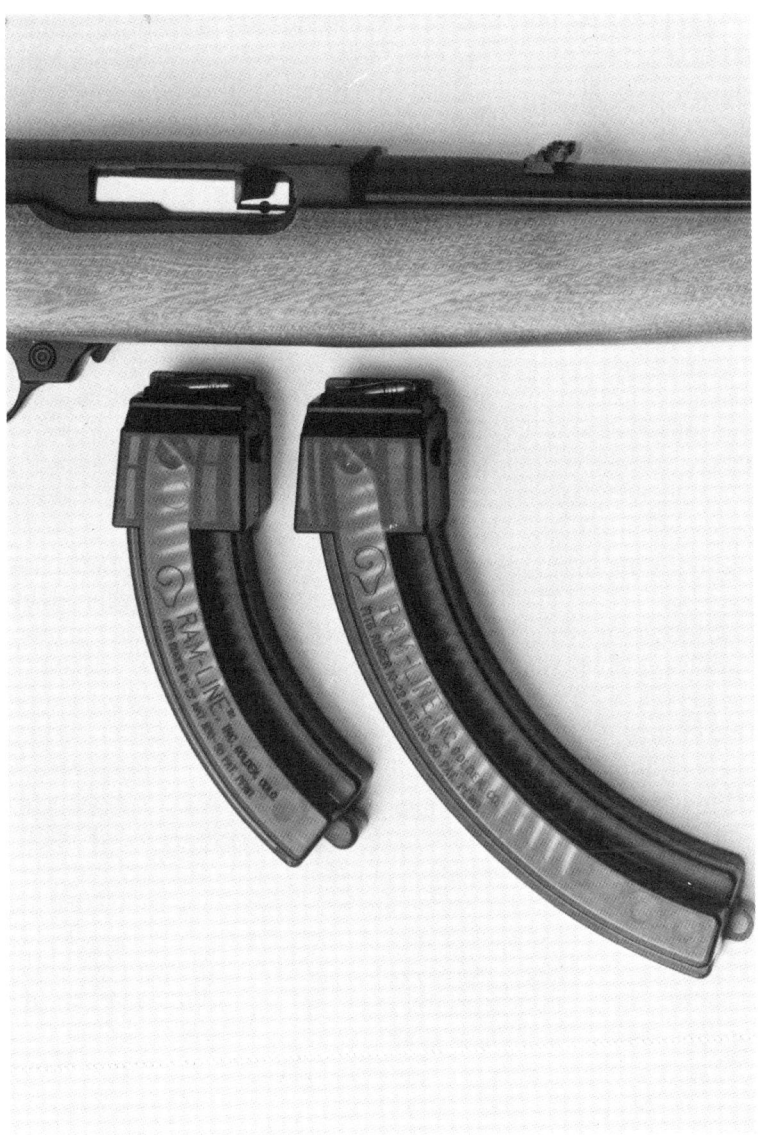

Ram-Line makes excellent extended "banana" magazines (the 30-round model is shown on the left and the 50-round model on the right). These magazines have a special pair of coiled springs, which keeps pressure quite low on the cartridges (and makes loading easy). Double-column cartridge storage in the magazine makes the length of the magazines relatively short. (Photo courtesy of Ram-Line)

in it, which makes loading the magazine less than thumb-busting. The Ram-Line magazines also use a double-column storage system, making them much shorter than magazines that use single-column stacking. Ram-Line's 50-round magazine is actually shorter than other companies' 25- or 30-round magazines.

The Ram-Line magazines are made with little holes and nubs so that they can be "jungle clipped" together. Care must be taken to avoid getting dirt in those magazines not yet inserted in the magazine well, since the feed lips are exposed. Firing from the prone position or dropping the rifle almost guarantees getting the open lips full of grime or even damaging the magazine when the jungle-clip arrangement is used. The Ram-Line 50-round magazines are slightly less prone to getting dirt in them with this arrangement, because they curve to the point that the spare magazine's lips face forward rather than downward.

Ram-Line offers its magazines in two types of polymer plastic: clear, which allows the shooter to see how many rounds are in a magazine, and black, which some shooters may prefer. Retail cost is about $39. for the 50-round magazine and about $18 for the black 30-round magazine, with the clear 30-round magazine costing $2 more.

Loading Ram-Line's extended magazines can be a quick job if you purchase the company's ALR-4300 Autoloader. This strange little gadget automatically aligns shells dumped into it from a box when you shake it. Once the shells are lined up, you clamp a Ram-Line magazine into the base and then load the magazine by pushing a rod in and out. The shells quickly click-clack into the magazine once you get used to the procedure. Hint: don't try to build up too much speed; make strokes steady and deliberate. Retail cost is about $40.

The Bingham all-steel, 30-round magazine is available from L.L. Baston Company. This magazine uses a single-column feed, making it a bit long, and requires a special adapter to fit into the wide magazine well of the 10/22. Cost for the magazine and adapter is approximately $20, while extra magazines cost about $13 each.

The Condor 25-round magazine was the first extended magazine commercially offered for the 10/22 and is still available from a number of mail-order companies as well as many gun

Accessories

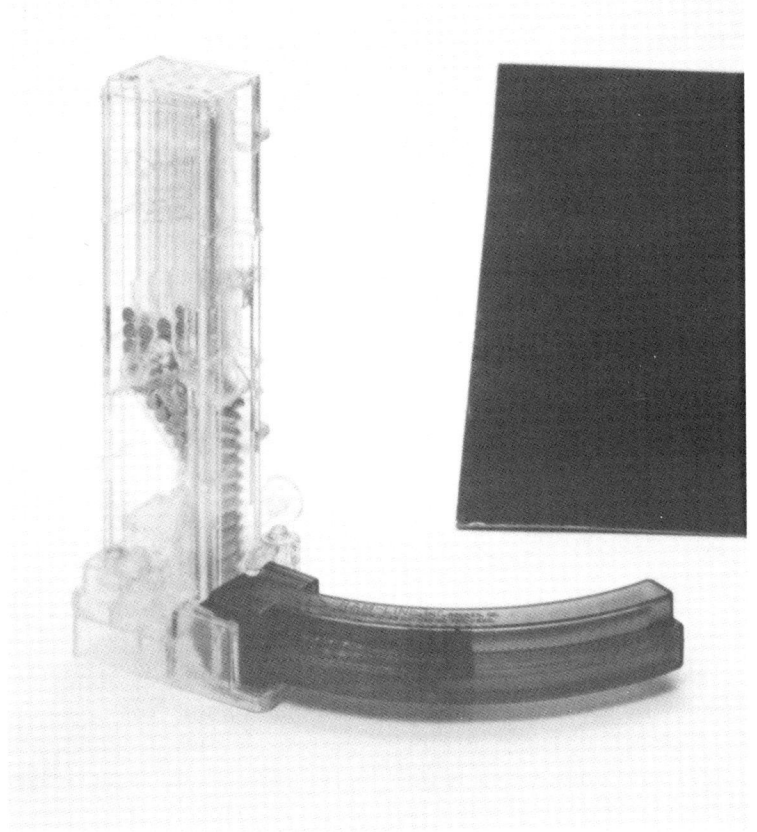

When shaken, Ram-Line's ALR-4300 Autoloader automatically aligns shells dumped into it from a box. Once the shells are lined up, clamping a Ram-Line magazine into the gadget's base allows the magazine to be quickly loaded by pushing a rod in and out. (Photo courtesy of Ram-Line)

stores. Unfortunately, high-tech advancements on the part of Ram-Line have overtaken the Condor magazine; being a single-column magazine, the old design is also rather long compared to the newer magazines. Eaton also makes a similar 25/22 magazine that offers little not found with others.

Mitchell Arms markets a 50/22 magazine that holds 50 rounds of ammunition in its tear-shaped body. The Mitchell magazine is loaded by removing its back and placing rounds into slots. Next, a spring is wound up, which pushes the string of shells around the magazine and up to the metal feed lips. The magazine also has a belt clip on its front so that it can be easily carried. Cost is about $40, with a choice of clear or black magazines. Both have clear back covers so that the user can tell at a glance how many rounds are in them.

Feather Enterprises offers a round, spring-powered magazine, the Sandford 50, which looks like the old Thompson submachine gun magazine. This magazine is a bit more high tech than the old magazines, however, and is made of plastic with a clear back.

E & L Manufacturing offers Dual and Quad magazine clips that allow standard Ruger 10-round magazines to be clipped together. The Quad version also has a storage compartment in its center for extra shells. While the Dual and Quad are of dubious use to many shooters, they may appeal to some, especially those who do not wish to use any magazines other than those made by Ruger. Cost is about $5 for the Dual and $10 for the Quad.

For those who dislike the standard magazine release on the 10/22, Ram-Line offers a plastic Kwik Release, which requires no gunsmithing work to replace the standard release. The unit extends downward slightly, in a manner similar to that of the M14 or Mini-14, and is simply pushed forward to release the magazine. This makes it possible to grasp an extended magazine and release it with one hand. Cost is about $4.

PELLET GUNS

If you think of pellet and BB guns as kid stuff, you probably haven't seen some of the newer versions of these guns. While the "European Sporter" versions of pellet guns are expensive propositions, this isn't true with some of the newer American-made guns, which make use of plastics, sheet metal stampings, and metal castings coupled with good engineering to produce high-powered, accurate rifles. These make the guns useful for honing target skills without burning up a lot of money. In addition, while buying .22 LR ammunition has probably never sent anyone to the poorhouse, there are times when the round is just too noisy or even dangerous to use for practice or plinking. In such a case, an air rifle can be the answer.

First choice in the inexpensive, American-made category is the Daisy Power Line 856. The gun is a "pump up," which means you can control the power of the pellets being shot from it. Velocity of pellets fired from the 856 can range from 295 to 630 feet per second (fps) according to the number of pumps used, from two to ten. The rifle's low velocity allows quiet indoor or backyard practice with a proper backstop. Maximum velocity obtained from ten pumps and lead pellets rather than BBs makes it possible to use the gun for hunting small game or varminting.

The 856's reservoir holds one hundred BBs; it can be used with lead pellets as well. By tilting the rifle back when pulling back on the loading port lever, a BB can be fed into the gun; tilting the 856 forward to keep a BB from feeding into the loading port allows a lead pellet to be placed in the loading port. This feature allows a shooter to switch from one to the other ammunition without unloading.

Steel BBs aren't as accurate as lead pellets and tend to ricochet if care isn't taken to use a proper backstop. But BBs are also as lethal at close ranges to small pests and don't contaminate an area with lead. They're also dirt cheap, making practice as close to free as a shooter will ever get. BB velocity for two to ten pumps runs from 315 to 650 fps.

The 856 has a mount for a .22 rifle scope molded into the top of its receiver. The low recoil of the 856 allows such scopes to be used without damaging the optical system and greatly improves its accuracy potential. Cost for the 856 is about $30.

The Daisy Power Line 922 is another inexpensive air rifle that is similar to the 856. The major differences are its caliber (.22), a slightly different charging handle, a 5-shot clip that allows pellets to be quickly fed into the chamber, and a Monte Carlo stock. Velocity with .22 lead pellets is from 270 to 530 fps with two to ten pumps.

REPLACEMENT SIGHTS

The front sight on the 10/22 is excellent and it's a little hard to imagine why it would need to be replaced. However, some shooters prefer glow-orange or white-blade front sights; Millet makes such sights, which are readily available from L.L. Baston for about $78.

L.L. Baston also offers Millet's rear sight for the 10/22, which has a square notch outlined in white. The big plus of this rear sight is that it is adjustable for windage without drifting the sight. It is possible to purchase the $57 replacement sight with either glow-orange or white outline colors.

Williams offers two aperture sights for the 10/22 as well as the .44 Carbine. These mount on the receiver using the scope mount holes drilled in the carbine or its scope mounting rails. In addition to giving the shooter the faster-to-use peep sight, the longer sighting radius of the Williams sights gives a greater degree of accuracy.

Ram-Line's Lookalike Sights have an AR-15 styling with a rear peep sight/carrying handle and elevated front sight. The rear sight assembly goes on the 10/22 scope mount, while the front assembly slips over the original front sight. In addition to the peep sights, it is also possible to mount a scope on the rear sight/carrying handle since it has scope rails molded into it. (Photo courtesy of Ram-Line)

Williams' Foolproof gives the finest adjustments, while the 5D Economy is nearly as precise. Cost of the Foolproof is about $35, while the 5D Economy costs around $20. Both are available from Brownells.

The Lyman 66 Receiver Sight is similar to the Williams sight but uses an audible-click, 1/4-minute adjustment system rather than a continuously variable, micrometer-style adjustment. Also available from Brownells, the Lyman 66 comes with two sight aperture inserts and costs about $46.

Feather Enterprises makes a front/rear sight kit that comes with the rear aperture sight incorporated into a scope mount. The new sights can be used even with the scope in place. The front sight consists of a ringed bar. The kit costs about $60; a rear/front-sight-only kit is also available for around $30 and a scope-mount-only kit for about $40.

Ram-Line offers a Lookalike Sights assembly for the 10/22 for those who want an AR-15-style rear peep sight/carrying handle and elevated front sight. The rear sight assembly goes on the 10/22 scope mount and is held in place with two screws, while the front assembly slips over the original front sight. In addition to the peep sights, it is also possible to mount a scope on the rear sight/carrying handle. The Lookalike Sights are a bit high for the standard Ruger stock but are just right for a proper cheek weld for most people when they use the Ram-Line folding stocks. Cost of the Lookalike Sights kit is about $20.

SCOPES AND MOUNTS

The 10/22 and .44 Carbine are accurate enough to merit good scopes. This is especially true with the 10/22, which cannot deliver its full potential with the cheap scopes that often end up on .22 rifles.

Scope mounting on either of these two carbines is simple. While the 10/22 base is set up to take 3/4-inch .22 scopes, scope rings that allow the mounting of 1-inch rifle scopes are readily available at most gun shops. With these rings, putting a good rifle scope in place is no problem. The 10/22 should come with a scope mount bar; if it is missing or lost, it is possible to purchase a new one from Sturm, Ruger and Company, Brownells, or other companies.

On the .44 Carbine, recoil can be a problem, so be sure always to check the mount for play. When mounting the scope on a .44 Carbine, it is also wise to place a little fine sandpaper grit between the rings and the scope and to use Loctite on any screws that may come loose on the base.

Getting a cheek weld can be a problem if you have high rings or rings that allow the iron sights to be used below the scope. Unless a large varminter scope or the like is mounted on the carbine, it's generally better to avoid high or see-through rings.

Modern scopes are better and tougher than their counterparts of a few years back. While you still can't plan on driving tent stakes with a quality scope, it will hold up with a little babying on your part. Range-finding scopes and other variable scopes are still relatively fragile when compared to fixed scopes and of doubtful use unless you're doing a lot of varmint work. Another plus of the fixed scope is that most of your cash goes toward optics rather than gear work. Of the fixed scopes, the 4x seems to be the best compromise for most shooters. Bausch and Lomb, Beeman, Bushnell, Leupold, Redfield, Shepherd, Simmons, Swarovski Optik, Tasco, Weatherby, Weaver, and Williams all have pretty good fixed scopes. Avoid unknown bargain scopes; they are seldom bargains.

Non-variable scopes are generally advertised with two numbers (for example, "4x40mm"). The first number is the power of the scope, while the second gives the size of the field of view. A wider field of view makes it possible to locate your target more easily. Unfortunately, many wide view scopes actually achieve the feeling of a wide view by cutting off part of the top and bottom of the view while the width remains the same as in comparable standard scopes. Save your money and get a round view scope with the same size field of view.

Small combat scopes may be ideal for carbines that will be used in the brush a lot. The best of these are Beeman's SS-1 and SS-2, which cost about $120 and $190 respectively. About the only drawback with these scopes is that they have a narrow field of view.

Many shooters find dot scopes, which place a dot of light in the field of view, quicker to use. Another plus is that many of these scopes are only 1x power so that the shooter can use both eyes when aiming. This cuts the time needed to acquire a moving target and allows the shooter a full field of view, two important considerations for hunting.

One style of dot scope is the occluded-eye scope, which creates a dot without any view of the target. The secret to using these scopes is to keep both eyes open. When this is done, the shooter's brain combines the pictures from each eye so that the dot seen by one eye seems to float over the scene seen by the other eye. This system is very quick and gives a wide field of view. While a few people have vision problems that make the occluded dot scope less than accurate, most people with good vision find it quick to use once they get over any squinting habits.

The best occluded-eye scope is the Armson O.E.G. (Occluded Eye Gunsight), which costs about $220 for the full-size version with a glow-in-the-dark insert. A smaller .22 version is also available; this has mounting rails for the standard .22 rifle mounting base like that of the 10/22, making it easy to install. These smaller versions cost about $100 with glow-in-the-dark inserts or around $75 for the daylight-only model. All the Armson O.E.G.s have a glow-red dot that is small enough not to cover distant targets but easy to find in most backgrounds.

The bargain-basement dot scope has to be Daisy's Point Aim scope. This little dot scope uses available light but is made for use with both eyes open (or with just the master eye open, as when using a regular scope). The shooter has a choice of a number of different reticles, including dots, rings, and cross hairs, which are easily mounted in place with tweezers. The scope is made of tough plastic and has .22-size scope mounts molded into its base so that it is easy to install. Best of all is the Point Aim's price tag, which is generally around $15.

A number of dot scopes create electronic dots that are superimposed in the shooter's view through the scope. The catch with these is that the user needs to replace the batteries from time to time and the brightness of the electric dot needs to be adjusted manually to match background brightness.

An electric dot scope does have some advantages over an available light scope: it can be used with just one eye, like a standard scope, and it can produce a much brighter dot at night than can the glowing element in the Armson scopes. The catch to nighttime use is that the human eye sees better from the side at night due to the arrangement of cone and rod cells at the back of the retina. This means that when something is viewed head-on in the dark, it may vanish from view, while if it is seen from the corner of the eye, it will become apparent. This is especially a problem if the scope has a bright dot in the center of view. Even if the dot is bright enough to show up, the object being viewed often vanishes in low light conditions. Nighttime scope use is not all one might hope for, although using one does help out in twilight shooting or other low light conditions indoors.

The two best-known brands of electric dot scopes are the Aimpoint, which costs around $180, and the Tascorama or Battery Dot Sight from Tasco, which sells for about $240. These sights use small batteries that power them for up to thousands of hours at the lower settings; brighter settings cause the batteries to fail sooner (be sure not to leave these on when you put them away). On both brands, the brightness is adjusted with a manually controlled rheostat. The Aimpoint 2000 is especially easy to mount on the 10/22 or .44 Carbine since the scope fits in standard 1-inch rings; a larger model of the Aimpoint is also available with a 3x scope attachment.

Some optical rifle scopes also use a glowing dot arrangement for twilight hunting use. One of these is the Bushnell Banner Lite-Site, which is normally used with the scope's cross hairs, but which can be switched on to place a glowing red dot in the center of the cross hairs.

Those worried about battery failure will be interested in the Armson Trijicon scope. It has cross hairs and a dot in its center; in daylight these aiming points are black, while at night the lines and dot take on a red glow, created by a very low-level, and very safe, radioactive source sealed in the scope. This makes it possible to readily see the cross hairs over the target in any type of light.

Several styles of Trijicon scopes are available, including a variable scope; one of the best is the 6x56mm, which has a very wide field of view as well as the ability to gather light, making it easier to see in dim conditions. While the scope is rather large, the trade-off is worth it. Cost is about $300.

Accessories 77

SILENCERS

The standard .22 LR cartridge is ideally suited to use with a sound suppressor, as most .22 bullets move at subsonic speeds and create only a small muzzle blast even when unsilenced. This is not the case with supersonic bullets like the new high-velocity or hyper-velocity rounds; supersonic bullets make a cracking sound, but these sounds will not be recognized as coming from a bullet by many people. Therefore, for those wishing only to disguise the sound of the firearm, rather than reduce its sound to a minimum, high-velocity .22 rounds can also be fired in a silencer-equipped 10/22.

While the screw-on silencer is ideal on pistols, such a device isn't satisfactory on a .22 rifle. Such a silencer often tends to be noisier at the end of a long barrel than it is on a short barrel. A better bet is the integral silencer, which is actually either a short barrel or a barrel with a number of holes drilled in it with a silencer built around it so that it appears to be a larger-than-normal barrel. These units can be quite soft and would certainly be ideal for those who wish to practice without alarming people or for some types of police work. For a good demonstration of such a weapon, see Anite Production's *Deadly Weapons* videocassette tape.

Jonathan Arthur Ciener can add an integral silencer to a 10/22. The finished unit looks like a "bull barrel" rifle; standard-velocity bullets leave the gun with very little noise. (Photo courtesy of Jonathan Arthur Ciener)

For those wishing more information about silencers, Paladin Press offers a number of good books. One to start with is *Silencers in the 1980s* for about $12. Remember, however, that making a silencer without prior approval is a federal offense; don't risk losing your right to own the 10/22 (or any other firearm) by trying to make a do-it-yourself illegal silencer.

Jonathan Arthur Ciener, Inc., is one of the best sources of commercial silencers and should be consulted if you decide you need to have a 10/22 modified by the addition of an integral silencer.

SLINGS

While most aftermarket stocks come with sling swivels, the standard .44 Carbine and 10/22 come without sling swivels. This isn't a problem, however, since swivels are easily mounted. A quick trip to the nearest gun store will generally reward the searcher with a set of swivels that screw into a wooden stock.

A bit more expensive, but well worth it, are Uncle Mike's detachable swivels. When a sling is not in use or the swivels not needed, a quick tug on the spring-loaded bar at one side of the swivels allows it to be rotated open and then popped off with the sling. All that's left behind is a rounded metal stud where the swing swivel was.

A huge array of slings is available. Again, the best--and usually the least expensive--slings are Uncle Mike's. This company's nylon slings are available in camouflage, brown, black, and white, come with padded "shoulder savers," and last almost forever. Another good nylon sling offered by Uncle Mike's is the Quick-Adjust, which can be quickly shortened so that it doesn't hang in the way when you don't need it.

Leather slings look nice but often have problems in rainy or damp environments. Many also promote rust if you leave your carbine in the back of a trunk or the like for a while with the leather resting against the barrel. Nylon slings make a lot more sense in this day and age.

Accessories

STOCKS

A number of companies offer plastic stocks for the 10/22. These are all easily installed by simply taking out the screws holding the standard stock in place, placing the action and barrel into the new stock, and screwing the stock onto the action.

The best assault-style stocks are made by Ram-Line and Choate Machine and Tool, but others may appeal to some shooters. Many of these aftermarket stocks are made of tough plastics, which are more resistant to abuse and moisture than wooden stocks. The flip side to using a plastic stock is that many ill-informed people will think you're carrying a machine gun or assault rifle.

For backpacking or storing a 10/22 in a vehicle, a folding stock, coupled with the Ram-Line Take-Down kit (see the next section), allows it to be stored in a very small space. Choate Machine and Tool offers a pistol-grip folding stock made of tough Zytel plastic with the positive steel lock-up mechanism used on the company's shotgun and rifle stocks. This makes a very strong but also somewhat heavy stock. The Choate stocks are quickly unfolded by pushing down on a small thumb button

Choate stocks are quickly unfolded or folded by pushing down on a small thumb button at the top of the pivot point. This action releases the stock so that it can be automatically locked open or closed. A barrel band allows the ring to be easily removed. (Photo courtesy of Choate Machine and Tool)

at the top of the pivot point; this releases the stock so that it can be swung out to lock automatically. A rubber butt pad, which helps hold the stock in place on the shoulder, is included on the Choate folder. Choate includes a metal spring and barrel ring, which replace the screw-held barrel band that comes with the 10/22. The Choate band is similar to that of the M1 Carbine and allows the ring to be removed by just depressing the spring on the stock rather than by having to loosen a screw. Choate is currently covering its metal stocks with plastic sheaths, which keep the metal from sticking to the skin in cold weather. This model of stock costs around $68.

In order to keep up with the competition, Choate also offers an economy folding stock made of the slightly less strong ABS plastic; it's nearly identical to the Zytel folder except for the lack of a rubber butt pad. Cost is about $50.

Choate started business by cutting off 10/22 stocks and placing the folding assembly on the 10/22s for the gun's owners. This type of work is still done by Choate Machine and Tool on a custom basis for anyone who desires the wooden stock rather than the plastic on his folder. However, most shooters prefer keeping the original stock so that they can change back to the factory configuration when they desire to do so.

Choate Machine and Tool's economy pistol-grip folding stock made of ABS plastic has the same positive steel lock-up mechanism used on the company's other folding stocks. This rifle also has Choate's handguard mounted over the barrel. (Photo courtesy of Choate Machine and Tool)

Parellex and Federal Ordnance offer wooden stocks coupled with metal stock assemblies that fold forward under the rifle's fore grip. The buttplate consists of an oval ring of metal; the metal stocks can be somewhat uncomfortable in cold weather. The stocks are unlocked by pushing on both sides of the swivel point over the pistol grips. Federal Ordnance has a third flat rod that runs alongside the two stock rods to keep the buttplate in the proper position when the stock is extended.

Feather Enterprises makes a wooden stock, the Falcon Foldout, which is similar to Sturm, Ruger and Company's folding stock sold for its Mini-14s. The Feather stock has a plastic pistol grip and a metal tube that forms the stock with a metal buttplate. Cost is about $98. For a time, Ram-Line also offered a wooden stock for the 10/22 that was modeled after the Ruger factory's Mini-14; these are still available on a custom basis.

Ram-Line's current production folder has done away with nearly all metal. This all-plastic, glass-filled stock is quite strong but has a bit more flex to it than does either a metal and plastic or wooden stock. The plus is that it's as light as the stan-

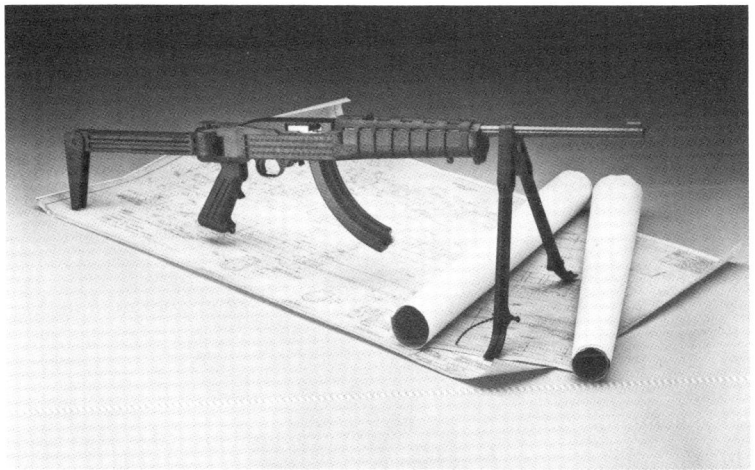

Ram-Line's glass-filled plastic folding stock has done away with the use of nearly all metal parts normally used in folding stocks. The big plus of this stock is that it's as light as the standard Ruger wooden stock; the weight of the rifle doesn't increase when the folding stock is used. The rifle is shown here with the Ram-Line plastic "clothespin" bipod, 50-round magazine, extended magazine release, and upper handguard. (Photo courtesy of Ram-Line)

dard Ruger wooden stock. The Ram-Line stock is ideal for use with a scope, as its straight-line configuration offers a cheek weld a bit higher than that of a wooden stock; this also makes it ideal for use with Ram-Line's Lookalike sight assembly.

The Ram-Line stock folds at the rear of the pistol grip and at the top of the buttplate. Large plastic buttons make it easy to fold or unfold. Folding the gun completely makes it a very compact package, though some may wish to leave the buttplate extended so that the stock can be brought into its extended position more quickly.

Ram-Line mounts its Laser AR-15 pistol grip on the Ram-Line stock. This grip has a slide-out plate on its bottom that allows the grip to be used as a storage compartment for small gear or ammunition. The grip has been placed back to give a longer space between the grip and trigger than is found with other stocks. For a shorter pull, it is easy to unscrew the grip, rasp out the forward end of the area on the stock where the pistol grip is

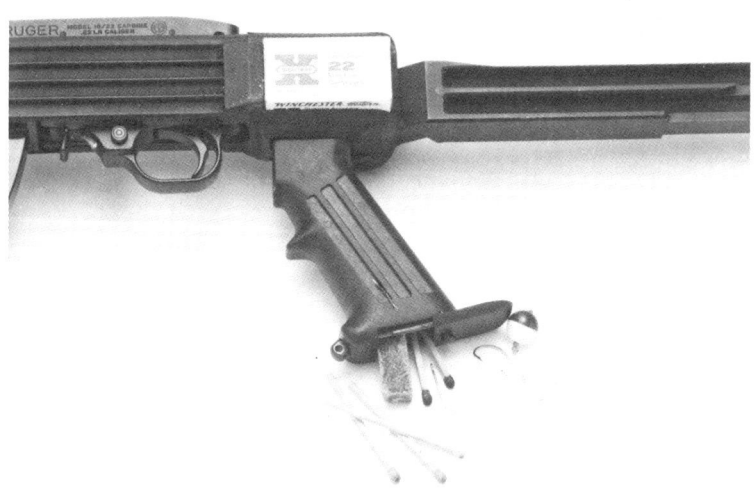

Ram-Line's folder sports the company's Laser AR-15 pistol grip, which has a trap door on its bottom. This allows it to be used as a storage compartment for small gear. A hollow area molded into the stock just above the grip will hold a spare box of .22 ammunition. (Photo courtesy of Ram-Line)

mounted, drill a new hole in the grip at its rear, and screw the grip back into place. Cost of the Ram-Line folder is about $68.

Choate offers stocks similar to its 10/22 stocks for the Mini-14 and M1 Carbine, while Ram-Line offers stocks similar to its Mini-14, Marlin 9mm and .45 ACP Camp Rifles, and the Marlin M-70. This may be a consideration for those wishing to create a family of similar rifles.

While aftermarket fixed stocks don't store in any shorter space than the standard Ruger wooden stock, they do offer a lot of resistance to abuse and moisture and also sport a pistol grip. Choate makes a fixed, pistol-grip plastic stock that is lighter than its folders, weighing about the same as the standard Ruger wooden stock. Like Choate's folders, the fixed stock sports a rubber buttplate and a spring-locked barrel band. Those finding the length of pull on the 10/22 a bit short will appreciate the Choate stock since it increases the pull length by three quarters of an inch. Cost is about $45.

Both Ram-Line and Choate offer ventilated handguards that go over the standard Ruger stock as well as fitting each company's stocks and matching them in style. While the need for a handguard on a .22 is not great, a handguard does help keep rust from developing on the barrel where the shooter's fingers touch it and also make the rifles look a bit more complete. Cost is about $15 for the Ram-Line model and about $8 for the Choate version.

Those who choose to stick with a wooden stock may wish to replace the plastic buttplate with a steel one similar to that used on the early Ruger carbines. Such a plate is available from Blake Tools for about $13.50. The checkering on this steel buttplate helps it hold its position better once in place.

Perhaps the ultimate in aftermarket stocks is the Simpson Arms Mannlicher stock kit, which is the spitting image of the original International offered by Ruger. The stock, which costs about $100, is made of walnut and is sold in an unfinished form so that the purchaser can finish it to his own taste. A thumb hole version of the stock is about $5 extra.

TAKEDOWN KITS

Ram-Line currently sells a Take-Down kit for about $27 that makes the 10/22 into a takedown rifle similar to the Marlin Papoose or AR-7. The Ram-Line kit replaces the stock and barrel band screws with knurled plastic knobs, which allow the gun to be fieldstripped without tools, and the 10/22's barrel block is replaced with one that has a knurled knob, making it possible to easily remove the barrel from the receiver without tools. With these changes, twisting a few knobs makes it possible to quickly assemble or disassemble a 10/22 and to store it in a very small space (especially with the Ram-Line folding stock), as well as making it easy to clean the barrel from the chamber end.

While the Take-Down block won't fit in the standard Ruger wooden stock or aftermarket stocks made by companies other than Ram-Line, it is possible to outlet stocks just below the barrel/receiver area to make room for the block and knobs. This simple task requires only some work with a wood chisel; for those who don't want to tackle this job, the kit will fit perfectly into the Ram-Line folding stock without modification. For compact storage/carrying of the 10/22, Ram-Line also offers a carrying kit that will hold one of its 10/22 folding stocks, the 10/22 receiver, the barrel, and several banana magazines. Cost is about $25.

****

The 10/22 and .44 Carbine are lightweight rifles that are ideal for a wide range of uses. As we've seen, the 10/22 is capable of being modified almost endlessly, while the .44 is a bit more limited. While many accessories will help some shooters, the wrong ones can greatly degrade the gun's reliability or even turn it into a Walter Mitty monstrosity.

To become a good shooter, develop your skills at shooting rather than looking for the gadget that magically will make you an expert. The results of careful practice with a rifle can be very rewarding, and much cheaper than searching for the magical accessory that will make up for lack of shooting skill.

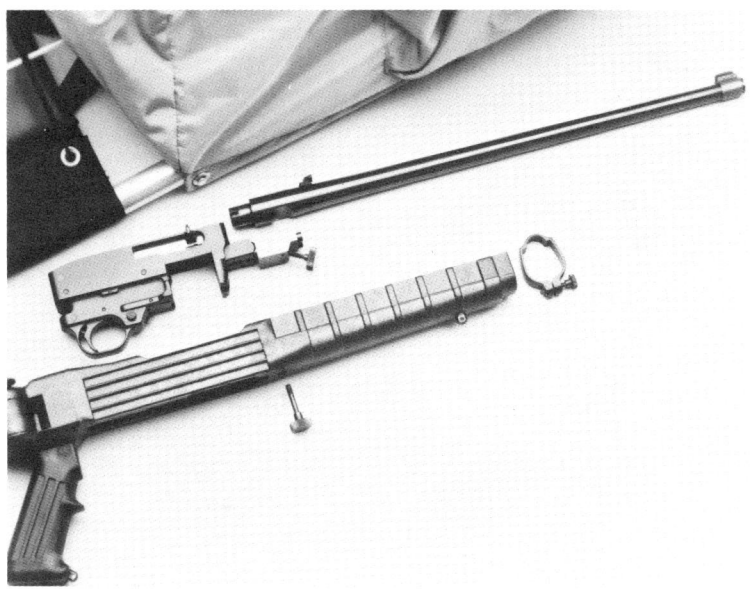

Ram-Line's Take-Down kit transforms the 10/22 into a takedown rifle similar to the Marlin Papoose or AR-7. The Ram-Line kit replaces the stock and barrel band screws with knurled plastic knobs (which allow the gun to be fieldstripped without a screwdriver), and the 10/22's barrel block is replaced with one having a knurled knob, making it possible to remove the barrel from the receiver without tools. (Photo courtesy of Ram-Line)

Chapter 5
Ammunition

Most 10/22 rifles seem to shoot tighter groups with some brands of ammunition than with others; unfortunately, this does not seem to be consistent from one rifle to another. In general, however, better ammunition gives better accuracy (among my favorite brands for the 10/22 are CCI, Federal, and Winchester). One trick used by some shooters is to weigh .22 ammunition; since brass is the most consistently made component in ammunition, when cartridges weigh more or less than their fellows, generally the bullet is a slightly different size or the powder charge isn't right. Removing these oddballs from the pack can shrink substantially the size of groups fired by a 10/22. Now let's look at some of the .22 rounds available for the 10/22.

.22 SHORT

The .22 Short is one of the oldest metallic cartridges, introduced in 1857. Although the round was originally designed for self-defense, it is less than adequate for that purpose. It is an ideal varmint round and will chamber in the 10/22, but it will not cycle the semiauto action and, with prolonged use, might cause chamber erosion.

The .22 Short is usually more expensive than the .22 LR and

offers little that the .22 LR doesn't. While the .22 Short's report is relatively soft compared to other rounds, the .22 CB Long Cap is even softer so that, here too, the .22 Short is second choice.

Velocity of the .22 Short varies substantially according to the barrel length of the weapon from which it is fired, with the maximum velocity coming from a barrel 6 to 7 inches long. A 2-inch barrel will have 846 fps; a 4-inch, 956 fps; and a 6-inch, 986 fps. An 8-inch may start to see a drop to 980 fps. That means that the 10/22 will actually be spitting out .22 Short bullets at velocities lower than those stated in factory charts.

If you must use a .22 Short, the CCI Mini-Mag Short HP might be first choice. Its much higher muzzle velocity gives it nearly twice the energy of standard .22 Short ammunition, and its hollow-point bullet makes it a bit more potent. Nevertheless, even with this top-line configuration, the .22 Short is not an ideal round for the 10/22.

.22 LONG

The .22 Long was developed in the late 1800s, at least a decade before the .22 LR. In power, the .22 Long is between the .22 Short and .22 LR with smokeless powder, though originally it had more power than either when all three used black powder. Like the .22 Short, the .22 Long will chamber in the 10/22 but it lacks the accuracy of the .22 Short and .22 LR. In addition, some rounds may push the bolt back slightly upon firing, which can create a jam with the empty brass caught between the bolt face and barrel.

.22 CB LONG

The .22 CB Cap was created as a low-velocity round for use in shooting galleries. CCI revived the idea as its .22 CB Long but changed the round from a cartridge that was the size of the .22 Short cartridge to that of the Long and Long Rifle, making it usable in guns chambered for either cartridge without chambering or chamber erosion problems.

The .22 CB Long makes it possible to shoot the 10/22 without ear protection; in fact, it allows nearly silent shooting, with the clatter of the action making more noise than the retort from the round. This makes the .22 CB Long ideal for training beginners, making nearly silent shots at varmints, or indoor practice with an adequate backstop.

While the round doesn't have enough power to cycle the 10/22's semiauto action, it is quite easy to chamber rounds by cycling the action by hand. This, too, can be good for training beginners, since it makes the accidental second shot that can occur with semiauto rifles in a beginner's hands nearly impossible.

The .22 CB Long changes a standard rifle into the ballistic equivalent of a super air gun, since the 29-grain bullet leaves the barrel at 727 fps. Taking into consideration the greater weight of the projectile, this makes the .22 with a .22 CB Long superior in power to all commercial air guns currently offered. Coupled with the low initial cost (you use the .22 rifle you already have) and the possibility of a quick follow-up shot (no cocking or pumping--just cycle the 10/22's action by hand), the air gun takes second place all the way around.

The only place the air rifle wins out is when it is inexpensive (which is not always the case, especially with expensive European sporters) and when a lot of ammunition is fired. When several thousand shots are made, the cost of the rimfire ammunition negates its benefits compared to the low cost of lead air-rifle pellets. Nevertheless, the flexibility offered in being able to make nearly silent shots from a 10/22 rifle that can also fire standard .22 LR ammunition is hard to beat.

.22 LR (LONG RIFLE)

The .22 LR was developed in 1887 and was originally loaded with black powder. The case on the standard .22 LR round is the same length as that of the .22 Long, but the bullet and overall case length are greater than those of the .22 Long.

At the turn of the century, the .22 LR made a successful transition to smokeless powder and became popular as a round for both pistols and rifles. Because of the immense popularity of the .22 LR and firearms chambered for it, a wealth of different types of ammunition is available. This gives the owner of a .22 LR firearm the ability to select ammunition that will suit a number of different purposes.

The .22 LR with high-velocity hollow points is ideal for small game within 75 yards; beyond that range, things get pretty chancy. That's not to say it isn't dangerous beyond that point; the little .22 bullet can zip out to a mile and be extremely dangerous for anyone it might chance to hit. To go this far, the bullet doesn't need a very high trajectory. Great care must be taken to have an adequate backstop when shooting .22s; while .22 rifles are a lot of fun, they aren't toys and should never be treated as such.

While best suited only to small game, the .22 LR can be an equally deadly round for medium-sized game (and humans). The catch is that it often does not kill quickly; animals (or people) often survive for several hours or even days before dying from internal bleeding when hit by a .22 LR. Because of this, the .22 LR is not an ideal round for larger targets, though it has been employed successfully both for self-defense and by poachers who use it at close ranges to make head shots on deer.

With Ram-Line and other companies now offering extended magazines that allow up to 50 rounds to be fired without reloading, one could argue that a 10/22 loaded with fast-velocity hollow points might be an effective self-defense weapon. Even if this is true, most people would certainly be better off with a larger caliber weapon. What it boils down to is that the .22 LR is marginal at best. If you must use the .22 LR for self-defense or for hunting medium-sized game, it can do the job. But there are a number of much better choices.

CCI's Stinger is probably the best round for hunting game at the large end of the "small game" scale (and the most suitable for use in a pinch for self-defense). The case of the Stinger cartridge is slightly longer than normal, allowing for a larger powder charge. The cartridge uses a slow-burning powder to give a greater velocity to its 32-grain bullet. This greater velocity makes the light bullet open up and expend its energy within most small game. This can also destroy a lot of meat, however, so for very small game, solid-point .22 bullets are sometimes better.

Other companies, including Remington, Federal, and Winchester, produce rounds similar to the Stinger. All of these rounds are marketed as "hyper-velocity" ammunition. They are hyper-velocity only when compared to other .22 LR ammunition; their speeds are still within the ranges of pistol bullets as opposed to the "high-velocity" designation normally used for rifle bullets traveling at speeds faster than 2,000 to 2,500 fps.

Because light bullets lose speed faster than heavier bullets, the Stinger and other "hyper" bullets are not as good as standard-weight .22 LR bullets beyond 80 to 100 yards. At this point, the heavier bullets have a higher speed, which translates into more power; however, most users of .22s won't use their firearms at such ranges; hunting should never be carried out beyond 100 yards with the .22 LR.

The second choice for hunting/defense use within 80 yards is the high-velocity .22 LR ammunition. Again, this is "high-velocity" only when compared to the older style .22 ammunition, and shouldn't be confused with high-velocity rifle ammunition. Ideally, this ammunition should have a hollow-point nose and a metallic jacket over the lead bullet to reduce barrel fouling.

The "standard" .22 LR round has a round lead bullet on it. This inexpensive ammunition is ideal for practice. Such ammunition is often less than ideal for hunting, however, since the bullets are a bit loose in the brass, and the powder and lubricants can quickly foul the 10/22. For practice or plinking, however, this round can offer a large savings and is often found at very low prices in discount stores.

.22 WINCHESTER MAGNUM RIMFIRE (WMR)

While Ruger doesn't offer a 10/22 chambered for this round, it is very possible that such a rifle might be offered in the near future by either Ruger or AMT.

The .22 WMR is a souped-up version of an old black-powder round, the WRF (Winchester Rimfire). The new .22 WMR round was introduced in 1959 by Winchester. In a rifle, the .22 WMR extends the useful range for hunting small game out to 100 to 125 yards. The catch is that the .22 WMR is overly destructive at close ranges, making its meat-gathering abilities less than ideal at close range.

Although the .22 LR will chamber in .22 WMR firearms, this isn't a safe practice, since chamber erosion can result and the brass will also consistently rupture, making a stuck case likely. Too, the .22 WMR bullet has a slightly larger diameter, making .22 LR bullets very inaccurate when fired from a .22 WMR gun.

Unfortunately, the .22 WMR isn't popular enough to merit a very large choice of ammunition in this chambering and it is also a bit more expensive than the .22 LR. The .22 Winchester Magnum Rimfire has a lot going for it ballistically, but it is still a second choice to the .22 LR for many purposes due to this lack of flexibility.

Federal's 50-grain HP (hollow-point) .22 WMR ammunition is the best for taking advantage of the extra range offered by the round. The bullet is ten grains heavier than other factory offerings, which gives the projectile more momentum at longer ranges. The soft copper jacket on the HP bullet expands somewhat, even at extreme ranges.

.44 SPECIAL

Although not often used in the .44 Carbine, the .44 Smith & Wesson Special will chamber in the firearm. This round was created in the early 1900s from the old black-powder .44 Russian cartridge to take advantage of the new smokeless powders. The round was introduced in 1907 by Smith & Wesson; both Colt and S&W made revolvers chambered for it. In the mid 1900s, Elmer Keith and others who were interested in using handguns for hunting created the .44 Magnum cartridge, which was actually a very hot .44 Special loading.

While the .44 Special is not powerful enough to cycle the .44 Carbine's semiauto action with any consistency, the round can be fed by hand. The .44 Special's recoil is considerably less, making it ideal for some types of training or practice; the round in its milder loads might be adapted to hunting small game as well.

.44 MAGNUM

The .44 Smith & Wesson Magnum was introduced in 1955, with Remington making the cartridge and Smith & Wesson offering a revolver chambered for the new round. Not long after, High Standard, Iver Johnson, Ruger, and others were making revolvers for the cartridge as well.

Because the .44 Magnum achieves its maximum velocity only with longer-barreled guns, and because of the loss of gas between a revolver's cylinder and barrel, it became obvious that a semiauto action could do much better ballistically than a revolver. With this in mind, Ruger introduced the .44 Carbine and quickly captured the market for a lightweight, short-range deer rifle that is very nearly the ballistic equivalent of the .30-30 cartridge.

While the .44 Magnum has proven too powerful for most people when used in a handgun, it is ideal in a carbine, offering good accuracy coupled with the penetration and power needed to quickly down a medium-sized animal; therefore, while the round has proven to be unpopular with most police forces, it is a nearly perfect hunting round for medium-sized game. Some brave souls in Alaska and elsewhere use this round for black bear, elk, moose, and brown bear; while it works at close range, many feel this round is marginal at best with these large and often dangerous animals.

Most of the major manufacturers offer excellent .44 Magnum ammunition. Lighter JHP (jacketed hollow-point) bullets generally give the best expansion, but the .44 Magnum is powerful enough at close range to practically guarantee expansion with lead and JSP (jacketed soft-point) bullets as well.

Appendix A
Manufacturers

Aimpoint USA
203 Elden St., Suite 302
Herndon, VA 22070
(Aimpoint electric dot scopes)

Arcadia Machine & Tool (AMT)
536 N. Vincent Ave.
Covina, CA 91722
(Manufacturer of the AMT Lightning pistol and 25/22 rifle)

Armson
P.O. Box 2130
Farmington Hills, MI 48018
(Armson O.E.G. available-light dot scope, glow-in-the-dark rifle and pistol sight inserts, and quality rifle scopes)

L.L. Baston Company
P.O. Box 1995
El Dorado, AR 71730
(Magazines, sights, and other firearms accessories)

Beeman Precision Arms
3440 Airway Drive
SantaRosa, CA 95403-2040
("Combat" scopes as well as air rifles, air pistols,
  and related accessories)

Blake Tools
Box 63-B
Park Ridge, NJ 07656-0063
(Manufacturer of 10/22 steel buttplate)

BMF Activator
3705 Broadway
Houston, TX 77017
(Fast-fire devices for the 10/22)

Brigade Quartermasters
1025 Cobb International Blvd.
Kennesaw, GA 30144
(Military surplus style equipment and accessories)

Brownells, Inc.
Route 2, Box 1
Montezuma, IA 50171
(Distributor of 10/22 and .44 Carbine replacement
  sights, scope mounts, and other accessories,
  as well as gunsmithing tools)

B-Square Company
P.O. Box 11281
Fort Worth, TX 76109
(Scope mounts for 10/22 and .44 Carbine)

Bushnell Optical Co.
2828 E. Foothill Blvd.
Pasadena, CA 91107
(Banner and other model scopes for rifles)

Manufacturers

Choate Machine and Tool Company
Box 218
Bald Knob, AR 72010
(Manufacturer of wide range of stocks, flash
 suppressors, handguards, and other accessories for the 10/22)

Jonathan Arthur Ciener, Inc.
6850 Riverside Drive
Titusville, FL 32780
(Manufacturer of silencers)

Daisy Manufacturing Company
P.O. Box 220
Rogers, AR 72757
(Makers of air guns suitable for practice)

E & L Manufacturing
Star Route 1, Box 569
Schoolhouse Road
Cave Creek, AZ 85331
(Barrel shrouds and magazine clips for 10/22)

Feather Enterprises
2300 Central Avenue
Boulder, CO 80301
(Accessories for the 10/22 and other firearms)

Federal Cartridge Corp.
2700 Foshay Tower
Minneapolis, MN 55402
(Manufacturer of .22 and .44 Magnum ammunition)

Federal Ordnance, Inc.
1443 Potrero Avenue
South El Monte, CA 91733
(10/22 folding stock, magazines, military surplus gear)

Firearm Systems and Design, Inc.
P.O. Box 19134
Minneapolis, MN 55419
(Manufacturer of "Ultimate" fast-fire attachment for 10/22)

Gun Parts (formerly Numrich Arms)
West Hurley, New York 12491
(Carries parts from some discontinued firearms)

Hansen Cartridge Company
244-246 Old Post Road
Southport, CT 06490
(Inexpensive .22 ammunition)

Harris Engineering
Barlow, KY 42024
(Harris bipod)

Michaels of Oregon ("Uncle Mike's")
P.O. Box 13010
Portland, OR 97213
(Rifle slings, scope covers, detachable sling swivels)

Millett Sights
16131 Gothard
Huntington Beach, CA 92647
(Rear replacement sights for 10/22 and .44 Carbine)

Mitchell Arms
19007 South Reyes Ave.
Compton, CA 90221
(Fast-fire accessories, magazines, and other accessories)

Olin
Winchester Group
120 Long Ridge Road
Stamford, CT 06904
(Winchester .22 and .44 Magnum ammunition)

Manufacturers

Omark Industries
P.O. Box 856
Lewiston, ID 83501
(Manufacturer of CCI .22 and .44 Magnum ammunition, as well as Weaver scopes and Outers shooting supplies)

Parellex Corporation
425 East Fourth Street
East Dundee, IL 60118
(Distributor of magazines, flash suppressors, slings, cases, scope mounts, etc., for 10/22)

Ram-Line, Inc.
15611 West 6th Street
Golden, CO 80401
(Manufacturer of 10/22 Laser plastic folding stocks; 10/22 Ultralight barrel, and other accessories for the 10/22 and other firearms)

SGW
624 Old Pacific Highway, South East
Olympia, WA 98503
(Manufacturer of AR-15 parts)

Shepherd Scope, Ltd.
Box 189
Waterloo, NE 68069
(Shepherd scopes)

Sherwood International
18714 Parthenia St.
Northridge, CA 91324
(Distributor of magazines, bayonets, slings, cases, etc.)

Sierra Supply
P.O. Box 1390
Durango, CO 81301
(Cleaning equipment, Break-Free CLP, and military ammunition/magazine carrying pouches as well as other surplus equipment)

Simpson Arms
1548D Adams Avenue
Costa Mesa, CA 92626
(Mannlicher stock kits for 10/22 carbine)

Sturm, Ruger & Company
Southport, CT 06490
(Manufacturers of the 10/22, .44 Carbine, and many other popular firearms)

Tasco Sales, Inc.
P.O. Box 520080
Miami, FL 33152
(Scopes and electric dot scope sights)

Appendix B

Publications & Videotapes

The following books, magazines, videos, and publishers provide valuable information about new products and developments with the 10/22 or other Ruger firearms, ammunition, and other related subjects.

American Rifleman magazine
1600 Rhode Island Avenue, NW
Washington, DC 20036

American Survival Guide magazine
McMullen Publishing
P.O. Box 15690
Santa Ana, CA 92705-0690

Conversations With Bill Ruger (videocassette tape)
Blacksmith Corp.
Box 424
Southport, CT 06490

Deadly Weapons (videocassette tape)
Anite Productions
P.O. Box 375
Pinole, CA 94564

Firepower magazine
Turbo Publishing, Inc.
P.O. Box 270
Cornville, AZ 86325

S.W.A.T. (Survival Weapons and Tactics) magazine
Turbo Publishing, Inc.
Cornville, AZ 86325

Troubleshooting Your Rifle and Shotgun
by J.B. Wood
DBI Books, Inc.
540 Frontage Road
Northfield, IL 60093